FUNCTIONAL CALCULUS

FUNCTIONAL CALCULUS

K. Kurien Issac
Ajeena Joseph
Renju Nandan

KNOWLEDGE
BAKERS

KNOWLEDGE BAKERS

ISBN: 978-93-49964-68-6

Published by : **Knowledge Bakers**
Pune, India
info@knowledgebakers.com

Preface

Functional Calculus provides an in-depth exploration of the theoretical foundations, methods, and applications of functional calculus, an area of mathematics that plays a central role in various branches of analysis, quantum mechanics, and differential equations. This book is designed for students, researchers, and professionals with an interest in functional analysis, operator theory, and mathematical physics, offering a comprehensive introduction to the concepts and tools of functional calculus and their real-world applications.

Functional calculus refers to the extension of traditional calculus concepts to more general settings, particularly when working with operators and functions defined on vector spaces, typically in Banach or Hilbert spaces. This field has found widespread applications in quantum mechanics, the study of differential operators, and the theory of partial differential equations. Through the methods of functional calculus, one can analyze and solve complex problems that involve linear operators, providing powerful tools to understand the behavior of physical systems modeled by such operators.

The book begins with a review of the essential mathematical foundations needed for the study of functional calculus, including the theory of normed spaces, operators, and spectral theory. It introduces readers to the fundamental concepts such as functions of operators, the spectrum of an operator, and various types of operator algebras that are crucial for understanding functional calculus in a modern mathematical context. Key tools, including the Borel functional calculus, Cauchy functional calculus, and the holomorphic functional calculus, are introduced and explored in detail.

Throughout the book, the emphasis is placed on understanding the interplay between abstract mathematical theory and its practical applications. Examples from quantum mechanics, such as the spectral theorem for self-adjoint operators, illustrate how functional calculus is applied to solve problems in physics. Furthermore, applications to the theory of partial differential equations and the study of dynamical systems highlight how these techniques are used to analyze and solve problems in mathematical physics and engineering.

A central theme of Functional Calculus is the study of different types of functions that can be applied to operators. The book provides a comprehensive treatment of operator-valued functions, spectral representations, and the use of functional calculus to compute operator functions in various settings, including bounded and unbounded operators. Special attention is given to the challenges posed by unbounded operators, which arise frequently in quantum mechanics and other physical theories.

In addition to operator theory, the book explores the relationship between functional calculus and other areas of mathematical analysis, such as harmonic analysis, Fourier analysis, and differential geometry. It addresses how functional calculus techniques can be extended to more general settings, such as non-commutative geometry and the study of Lie groups and algebras, offering readers a broader view of how functional calculus intersects with other advanced mathematical fields.

The book also provides a variety of exercises, worked examples, and case studies to help readers understand both the theoretical aspects of functional calculus and its practical uses. These exercises range from basic to advanced levels, allowing readers to apply the concepts they have learned and develop a deeper understanding of how functional calculus is used to solve real-world problems.

Functional Calculus serves as both a textbook and a reference for anyone interested in this powerful branch of mathematics. Whether you are a student seeking a deeper understanding of functional calculus or a researcher looking for a comprehensive guide to the applications of these techniques, this book offers the necessary theoretical background and practical tools to navigate this complex field.

We extend our thanks to the many contributors and researchers whose work has shaped the development of functional calculus and has contributed to the creation of this book. It is our hope that Functional Calculus will inspire continued exploration in this fascinating area of mathematics and contribute to the advancement of both theoretical research and practical applications in mathematics, physics, and engineering.

K. Kurien Issac

Ajeena Joseph

Renju Nandan

Contents

01 DETERMINANTAL REPRESENTATIONS OF THE CORE INVERSE AND ITS GENERALIZATIONS

1. INTRODUCTION: AN OVERVIEW

In the whole chapter, the notations $\mathbb{R}$ and $\mathbb{C}$ are reserved for fields of the real and complex numbers, respectively. $\mathbb{C}^{m\times n}$ stands for the set of all $m \times n$ matrices over $\mathbb{C}$. $\mathbb{C}_r^{m\times n}$ determines its subset of matrices with a rank *r*. For A$\in \mathbb{C}^{m\times n}$ the symbols A^* and rk(**A**) specify the conjugate transpose and the rank of A, respectively, $|A|$ or *det***A** stands for its determinant. A matrix A$\in \mathbb{C}^{n\times n}$ is Hermitian if $A^*=A$.

$A^\dagger$ means *the Moore-Penrose inverse* of A$\in\mathbb{C}^{n\times m}$,i.e., the exclusive matrix X satisfying the following four equations:

$$AXA = A \tag{1}$$

$$XAX = X \tag{2}$$

$$(AX)^* = AX \tag{3}$$

$$(XA)^* = XA \tag{4}$$

For A$\in \mathbb{C}^{n\times n}$ with index IndA=*k*, i.e., the smallest positive number such that $rk(A^{k+1})=rk(A^k)$, *the Drazin inverse* of **A**, denoted by A^d, is called the unique matrix **X** that satisfies Eq. (2) and the following equations,

$$AX = XA; \tag{5}$$

$$XA^{k+1} = A^k \tag{6}$$

$$A^{k+1}X = A^k. \tag{7}$$

In particular, if Ind **A**=1, then the matrix **X** is called *the group inverse*, and it is denoted by $X=A^\#$. If Ind A=0, then **A** is nonsingular and $A^d=A^\dagger=A^{-1}$.

It is evident that if the condition (5) is fulfilled, then (6) and (7) are equivalent. We put both these conditions because they will be used below independently of each other and without the obligatory fulfillment of (5).

A matrix **A** satisfying the conditions (*i*),(*j*),... is called an {*i,j*,...} inverse of **A**, and is denoted by $A^{(i,j,...)}$ The set of matrices $A^{(i,j,...)}$ is denoted A{*i,j*,...}. In particular, $A^{(1)}$ is called the inner inverse, $A^{(2)}$ is called the outer inverse, $A^{(1,2)}$ is called the reflexive inverse, $A^{(1,2,3,4)}$ is the Moore-Penrose inverse, etc.

For an arbitrary matrix $A\in\mathbb{C}^{m\times n}$, we denote by

- N (A) = $\{x\in\mathbb{H}^{n\times1} : Ax = 0\}$, the kernel (or the null space) of A;
- C (A) = $\{y\in\mathbb{H}^{m\times1} : y = Ax, x\in\mathbb{H}^{n\times1}\}$, the column space (or the range space) of A; and
- R (A) = $\{y\in\mathbb{H}^{1\times n} : y = xA, x\in\mathbb{H}^{1\times m}\}$, the row space of A.

$P_A := AA^{\dagger}$ and $Q_A := A^{\dagger}A$ are the orthogonal projectors onto the range of **A** and the range of A^*, respectively.

The core inverse was introduced by Baksalary and Trenkler in [1]. Later, it was investigated by S. Malik in [2] and S.Z. Xu et al. in [3], among others.

Definition 1.1. [1] A matrix $X\in\mathbb{C}^{n\times n}$ is called the core inverse of $\mathbf{A}\in\mathbb{C}^{n\times n}$ if it satisfies the conditions

$$\mathbf{AX} = \mathbf{P}_A, and\ C(\mathbf{X}) = C(\mathbf{A}).$$

When such matrix X exists, it is denoted as $A^{\oplus}$.

In 2014, the core inverse was extended to the core-EP inverse defined by K. Manjunatha Prasad and K.S. Mohana [4]. Other generalizations of the core inverse were recently introduced for $n \times nn \times n$ complex matrices, namely BT inverses [5], DMP inverses [2], CMP inverses [6], etc. The characterizations, computing methods, and some applications of the core inverse and its generalizations were recently investigated in complex matrices and rings (see, e.g., [7, 8, 9, 10, 11, 12, 13, 14, 15, 16, 17, 18]).

In contrast to the inverse matrix that has a definitely determinantal representation in terms of cofactors, for generalized inverse matrices, there exist different determinantal representations as a result of the search of their more applicable explicit expressions (see, e.g. [19, 20, 21, 22, 23, 24, 25]). In this chapter, we get new determinantal representations of the core inverse and its generalizations using recently obtained by the author determinantal representations of the Moore-Penrose inverse and the Drazin inverse over the quaternion skew field, and over the field of complex numbers as a special case [26, 27, 28, 29, 30, 31, 32, 33, 34]. Note that a determinantal representation of the core-EP generalized inverse in complex matrices has been derived in [4], based on the determinantal representation of an reflexive inverse obtained in [19, 20].

The chapter is organized as follows: in Section 2, we start with preliminary introduction of determinantal representations of the Moore-Penrose inverse and the Drazin inverse. In Section 3, we give determinantal representations of the core inverse and its generalizations, namely the right and left core inverses are established in Section 3.1, the core-EP inverses in Section 3.2, the core DMP inverse and its dual in Section 3.3, and finally the CMP inverse in Section 3.4. A numerical example to illustrate the main results is considered in Section 4. Finally, in Section 5, the conclusions are drawn.

2. PRELIMINARIES

Let a:$\{a_1,\ldots,a_k\} \subseteq \{1,\ldots,m\}$ and β: = $\{\beta_1,\ldots,\beta K\} \subseteq \{1,\ldots,n\}$ be subsets with $1\leq k\leq \min\{m,n\}$. By $\mathbf{A}^a{}_\beta$, we denote a submatrix of $A\in\mathbb{H}^{m\times n}$ with rows and columns indexed by α and β, respectively. Then, $A^a{}_a$ is a principal submatrix of A with rows and columns indexed by α, and $|A|^a{}_a|A|^{\alpha}_{\alpha}$ is the corresponding principal minor of the determinant |A|. Suppose that stands for the collection of strictly increasing sequences of $1\leq k\leq n1 \leq k \leq n$ integers chosen from $\{1,\ldots,n\}$. For fixed $i\in\alpha$ and $j\in\beta$, put $I_{r,m}\{i\}:=\{\alpha:\alpha\in L_{r,m}, i\in\alpha\}$ and $J_{r,n}\{j\}:=\{\beta:\beta\in L_{r,n}, j\in\beta\}$.

$$L_{k,n} := \{\alpha : \alpha = (\alpha_1, \ldots, \alpha_k), 1 \leq \alpha_1 < \cdots < \alpha_k \leq n\}$$

The *j*th columns and the *i*th rows of A and A^* denote $a_{.j}$ and $a^*_{.j}$ and $a_{i.}$ and $a^*_{i.}$, respectively. By Ai.(b) and A.j(c), we denote the matrices obtained from A by replacing its *i*th row with the row b, and its *j*th column with the column c.

Theorem 2.1. [28] *If, $A\in\mathbb{H}_r^{m\times n}$, then the Moore-Penrose inverse $\mathbf{A}^\dagger=(a_{ij}^\dagger a_{ij}^\dagger)\in\mathbb{C}^{n\times m}$ possesses the determinantal representations*

$$a_{ij}^{\dagger} = \frac{\sum_{\beta\in J_{r,n}\{i\}} \left|(\mathbf{A}^*\mathbf{A})_{.i}\left(\mathbf{a}^*_{.j}\right)\right|^{\beta}_{\beta}}{\sum_{\beta\in J_{r,n}} |\mathbf{A}^*\mathbf{A}|^{\beta}_{\beta}} = \tag{8}$$

$$= \frac{\sum_{\alpha\in I_{r,m}\{j\}} \left|(\mathbf{AA}^*)_{j.}\left(\mathbf{a}^*_{i.}\right)\right|^{\alpha}_{\alpha}}{\sum_{\alpha\in I_{r,m}} |\mathbf{AA}^*|^{\alpha}_{\alpha}}. \tag{9}$$

Remark 2.2. For an arbitrary full-rank matrix $A\in\mathbb{C}_r^{m\times n}$, a row vector $b\in\mathbb{H}^{1\times m}$, and a column-vector $c\in\mathbb{H}^{n\times 1}$, we put, respectively,

$$\left|(\mathbf{AA}^*)_{i.}(\mathbf{b})\right| = \sum_{\alpha\in I_{m,m}\{i\}} \left|(\mathbf{AA}^*)_{i.}(\mathbf{b})\right|^{\alpha}_{\alpha}, \quad i = 1, \ldots, m,$$

$$|\mathbf{AA}^*| = \sum_{\alpha\in I_{m,m}} |\mathbf{AA}^*|^{\alpha}_{\alpha}, \quad \text{when } r = m;$$

$$\left|(\mathbf{A}^*\mathbf{A})_{.j}(\mathbf{c})\right| = \sum_{\beta\in J_{n,n}\{j\}} \left|(\mathbf{A}^*\mathbf{A})_{.j}(\mathbf{c})\right|_{\beta}^{\beta}, \quad j=1,\dots,n,$$

$$|\mathbf{A}^*\mathbf{A}| = \sum_{\beta\in J_{n,n}} |\mathbf{A}^*\mathbf{A}|_{\beta}^{\beta}, \quad \text{when } r=n.$$

Corollary 2.3. [21] *Let* $\mathbf{A}\in\mathbb{C}_r^{m\times n}$. *Then, the following determinantal representations can be obtained*

(i). for the projector $Q_A=(q_{ij})_{n\times n}$,

$$q_{ij} = \frac{\sum_{\beta\in J_{r,n}\{i\}} \left|(\mathbf{A}^*\mathbf{A})_{.i}(\dot{\mathbf{a}}_{.j})\right|_{\beta}^{\beta}}{\sum_{\beta\in J_{r,n}} |\mathbf{A}^*\mathbf{A}|_{\beta}^{\beta}} = \frac{\sum_{\alpha\in I_{r,n}\{j\}} \left|(\mathbf{A}^*\mathbf{A})_{j.}(\dot{\mathbf{a}}_{i.})\right|_{\alpha}^{\alpha}}{\sum_{\alpha\in I_{r,n}} |\mathbf{A}^*\mathbf{A}|_{\alpha}^{\alpha}}, \tag{10}$$

where $\dot{\mathbf{a}}_{.j}$ is the jth column and $\dot{\mathbf{a}}_{i.}$ is the ith row of $\mathbf{A}^*\mathbf{A}$; and

(ii). for the projector $P_A=(p_{ij})_{m\times m}$,

$$p_{ij} = \frac{\sum_{\alpha\in I_{r,m}\{j\}} \left|(\mathbf{A}\mathbf{A}^*)_{j.}(\ddot{\mathbf{a}}_{i.})\right|_{\alpha}^{\alpha}}{\sum_{\alpha\in I_{r,m}} |\mathbf{A}\mathbf{A}^*|_{\alpha}^{\alpha}} = \frac{\sum_{\beta\in J_{r,m}\{i\}} \left|(\mathbf{A}\mathbf{A}^*)_{.i}(\ddot{\mathbf{a}}_{.j})\right|_{\beta}^{\beta}}{\sum_{\beta\in J_{r,m}} |\mathbf{A}\mathbf{A}^*|_{\beta}^{\beta}}, \tag{11}$$

where $\ddot{\mathbf{a}}_{i.}$ is the ith row and $\ddot{\mathbf{a}}_{.j}$ is the jth column of $\mathbf{A}\mathbf{A}^*$.

The following lemma gives determinantal representations of the Drazin inverse in complex matrices.

Lemma 2.4. [21] *Let* $\mathbf{A}\in\mathbb{C}^{n\times n}$ *with* Ind $A=k$ *and* rk $\mathbf{A}^{k+1}=$rk $\mathbf{A}^k=r$. *Then, the determinantal representations of the Drazin inverse* $A^d=(a^d_{ij})\in\mathbb{C}^{n\times n}$ *are*

$$a_{ij}^d = \frac{\sum_{\beta\in J_{r,n}\{i\}} \left|\left(\mathbf{A}^{k+1}\right)_{.i}\left(\mathbf{a}_{.j}^{(k)}\right)\right|_{\beta}^{\beta}}{\sum_{\beta\in J_{r,n}} \left|\mathbf{A}^{k+1}\right|_{\beta}^{\beta}} = \tag{12}$$

$$= \frac{\sum_{\alpha\in I_{r,n}\{j\}} \left|\left(\mathbf{A}^{k+1}\right)_{j.}\left(\mathbf{a}_{i.}^{(k)}\right)\right|_{\alpha}^{\alpha}}{\sum_{\alpha\in I_{r,n}} \left|\mathbf{A}^{k+1}\right|_{\alpha}^{\alpha}}, \tag{13}$$

Where $a_{i.}^{(k)}$ is the ith row and $a_j^{(k)}$ $a_{.j}^{(k)}$ is the jth column of $\mathbf{A}^k$.

Corollary 2.5. [21] *Let* $A\in\mathbb{C}^{n\times n}$ *with* Ind $A=1$ *and* rk$\mathbf{A}^2=$rk$\mathbf{A}=r$. *Then, the determinantal representations of the group inverse* $\mathbf{A}^{\#}=(a^{\#}_{ij})\in\mathbb{C}^{n\times n}$ *are*

$$a_{ij}^{\#} = \frac{\sum_{\beta \in J_{r,n}\{i\}} \left| \left(\mathbf{A}^2\right)_{.i} \left(\mathbf{a}_{.j}\right) \right|_{\beta}^{\beta}}{\sum_{\beta \in J_{r,n}} \left| \mathbf{A}^2 \right|_{\beta}^{\beta}} = \frac{\sum_{\alpha \in I_{r,n}\{j\}} \left| \left(\mathbf{A}^2\right)_{j.} \left(\mathbf{a}_{i.}\right) \right|_{\alpha}^{\alpha}}{\sum_{\alpha \in I_{r,n}} \left| \mathbf{A}^2 \right|_{\alpha}^{\alpha}}. \tag{14}$$

3. DETERMINANTAL REPRESENTATIONS OF THE CORE INVERSE AND ITS GENERALIZATIONS

3.1 Determinantal representations of the core inverses

In addition to the core inverse mentioned in [35], the dual core inverse was also intended to be introduced. Since both of these core inverses are equivalent and differ only in their position relative to the inducting matrix A, we recommend referring to them as the right and left core inverses based on their placements. According to reference [1], we have an alternative definition that is equivalent to Definition 1.1.

Definition 3.1. A matrix $X \in \mathbb{C}^{n\times n}$ is said to be the right core inverse of $A \in \mathbb{C}^{n\times n}$ if it satisfies the conditions

$$\mathbf{AX} = \mathbf{P}_A, and\ \mathcal{C}(\mathbf{X}) = \mathcal{C}(\mathbf{A}).$$

When such matrix X exists, it is denoted as $A^{\textcircled{\#}}$.

The following definition of the left core inverse can be given that is equivalent to the introduced dual core inverse [35].

Definition 3.2 A matrix $X \in \mathbb{C}^{n\times n}$ is said to be the left core inverse of $A \in \mathbb{C}^{n\times n}$ if it satisfies the conditions

$$\mathbf{XA} = \mathbf{Q}_A, and\ \mathcal{R}(\mathbf{X}) = \mathcal{R}(\mathbf{A}). \tag{15}$$

When such matrix X exists, it is denoted as $A_{\textcircled{\#}}$.

Remark 3.3. In [35], the conditions of the dual core inverse are given as follows:

$$A_{\textcircled{\#}}\mathbf{A} = \mathbf{P}_{A^*}, and\ \mathcal{C}(A_{\textcircled{\#}}) \subseteq \mathcal{C}(\mathbf{A}^*).$$

Since $\mathrm{P}_{A^*} = \mathrm{A}^*(\mathrm{A}^*)^{\dagger} = (\mathrm{A}^{\dagger}\mathrm{A})^* = \mathrm{A}^{\dagger}\mathrm{A} = \mathrm{Q}_A$, and $\mathcal{R}(\mathrm{A}) = \mathcal{C}(\mathrm{A}^*)$;, then these conditions and (15) are analogous.

Due to [1], we introduce the following sets of quaternion matrices

$$\mathbb{C}_n^{\mathrm{CM}} = \{\mathrm{A} \in \mathbb{C}^{n\times n} : \mathrm{rk}\,\mathrm{A}^2 = \mathrm{rk}\,\mathrm{A}\},$$

$$\mathbb{C}_n^{\mathrm{EP}} = \{\mathrm{A} \in \mathbb{C}^{n\times n} : \mathrm{A}^{\dagger}\mathrm{A} = \mathrm{A}\mathrm{A}^{\dagger}\} = \{\mathcal{C}(\mathrm{A}) = \mathcal{C}(\mathrm{A}^*)\}.$$

The matrices from $\mathbb{C}_n^{\mathrm{CM}}$ are called group matrices or core matrices. If $A \in \mathbb{C}_n^{\mathrm{EP}}$, then clearly $A^{\dagger} = A^{\#}$. It is known that the core inverses of $A \in \mathbb{C}^{n\times n}$ exist if and only if $A \in \mathbb{C}_n^{\mathrm{CM}}$ or Ind A=1. Moreover, if A is nonsingular, Ind **A**=0, then its core inverses are the usual inverse. Due to [1], we have the following representations of the right and left core inverses.

Lemma 3.4. [1] *Let* $\mathbf{A}\in\mathbb{C}_n^{CM}\cdot\mathbb{C}_n^{CM}$. *Then,*

$$\mathbf{A}^{\circledast} = \mathbf{A}^{\#}\mathbf{A}\mathbf{A}^{\dagger}, \tag{16}$$

$$\mathbf{A}_{\circledast} = \mathbf{A}^{\dagger}\mathbf{A}\mathbf{A}^{\#} \tag{17}$$

Remark 3.5. In Theorems 3.6 and 3.7, we will suppose that $\mathbf{A}\in\mathbb{C}_n^{CM}$ but $\mathbf{A}\notin\mathbb{C}_n^{EP}$. Because, if $\mathbf{A}\in\mathbb{C}_n^{CM}$ (in particular, **A** is Hermitian), then from Lemma 3.4 and the definitions of the Moore-Penrose and group inverses, it follows that $\mathbf{A}^{\circledast} = \mathbf{A}_{\circledast} = \mathbf{A}^{\#} = \mathbf{A}^{\dagger}$.

Theorem 3.6.*Let* $\mathbf{A}\in\mathbb{C}_n^{CM}$ *and* $\operatorname{rk}\mathbf{A}^2=\operatorname{rk}\mathbf{A}=s$. *Then, its right core inverse has the following determinantal representations*

$$a_{ij}^{\circledast,r} = \frac{\sum_{\alpha\in I_{s,n}\{j\}}\left|\left(\mathbf{A}\mathbf{A}^*\right)_{j.}\left(\mathbf{u}_{i.}^{(1)}\right)\right|_{\alpha}^{\alpha}}{\sum_{\beta\in J_{s,n}}\left|\mathbf{A}^2\right|_{\beta}^{\beta}\sum_{\alpha\in I_{s,n}}\left|\mathbf{A}\mathbf{A}^*\right|_{\alpha}^{\alpha}} = \tag{18}$$

$$= \frac{\sum_{\beta\in J_{s,n}\{i\}}\left|\left(\mathbf{A}^2\right)_{.i}\left(\mathbf{u}_{.j}^{(2)}\right)\right|_{\beta}^{\beta}}{\sum_{\beta\in J_{s,n}}\left|\mathbf{A}^2\right|_{\beta}^{\beta}\sum_{\alpha\in I_{s,n}}\left|\mathbf{A}\mathbf{A}^*\right|_{\alpha}^{\alpha}}, \tag{19}$$

where

$$\mathbf{u}_{i.}^{(1)} = \left[\sum_{\beta\in J_{s,n}\{i\}}\left|\left(\mathbf{A}^2\right)_{.i}\left(\tilde{\mathbf{a}}_{.f}\right)\right|_{\beta}^{\beta}\right]\in\mathbb{C}^{1\times n},\ f=1,\dots,n$$

$$\mathbf{u}_{.j}^{(2)} = \left[\sum_{\alpha\in I_{s,n}\{j\}}\left|\left(\mathbf{A}\mathbf{A}^*\right)_{j.}\left(\tilde{\mathbf{a}}_{l.}\right)\right|_{\alpha}^{\alpha}\right]\in\mathbb{C}^{n\times 1},\ l=1,\dots,n.$$

are the row and column vectors, respectively. Here $\tilde{\mathbf{a}}_{.f}$ and $\tilde{\mathbf{a}}_{l.}$ are the *f*th column and *l*th row of $\tilde{\mathbf{A}} := \mathbf{A}^2\mathbf{A}^*$.

Proof. Taking into account (16), we have for $^{\#}\mathbf{A}$,

$$a_{ij}^{\#,r} = \sum_{l=1}^{n}\sum_{f=1}^{n} a_{il}^{\#}a_{lf}a_{fj}^{\dagger}. \tag{20}$$

By substituting (14) and (15) in (20), we obtain

$$a_{ij}^{\#,r}=\sum_{l=1}^{n}\frac{\sum_{f=1}^{n}\sum_{\beta\in J_{s,n}\{i\}}\left|(\mathbf{A}^2)_{.i}(\mathbf{a}_{.f})\right|_{\beta}^{\beta}a_{fl}}{\sum_{\beta\in J_{s,n}}\left|\mathbf{A}^2\right|_{\beta}^{\beta}}\frac{\sum_{\alpha\in I_{s,n}\{j\}}\left|(\mathbf{AA}^*)_{j.}(\mathbf{a}_{l.}^*)\right|_{\alpha}^{\alpha}}{\sum_{\alpha\in I_{s,n}}\left|\mathbf{AA}^*\right|_{\alpha}^{\alpha}}=$$

$$\frac{\sum_{f=1}^{n}\sum_{l=1}^{n}\sum_{\beta\in J_{s,n}\{j\}}\left|(\mathbf{A}^2)_{.j}(\mathbf{e}_{.f})\right|_{\beta}^{\beta}\tilde{a}_{fl}\sum_{\alpha\in I_{s,n}\{j\}}\left|(\mathbf{AA}^*)_{j.}(\mathbf{e}_{l.})\right|_{\alpha}^{\alpha}}{\sum_{\beta\in J_{s,n}}\left|\mathbf{A}^2\right|_{\beta}^{\beta}\sum_{\alpha\in I_{s,n}}\left|\mathbf{AA}^*\right|_{\alpha}^{\alpha}},$$

where $\mathrm{e}_{.l}$ and $\mathrm{e}_{l.}$ are the unit column and row vectors, respective[illegible]h that all their components are 0, except the lth components which are 1; $\tilde{a}_{lf}$ is the (lf)th element of the matrix $\tilde{\mathbf{A}}:=\mathbf{A}^2\mathbf{A}^*$.

Let

$$u_{il}^{(1)}:=\sum_{f=1}^{n}\sum_{\beta\in J_{s,n}\{i\}}\left|(\mathbf{A}^2)_{.i}(\mathbf{e}_{.f})\right|_{\beta}^{\beta}\tilde{a}_{fl}=\sum_{\beta\in J_{s,n}\{i\}}\left|(\mathbf{A}^2)_{.i}(\tilde{\mathbf{a}}_{.l})\right|_{\beta}^{\beta},\quad i,l=1,\dots,n.$$

Construct the matrix $\mathrm{U}_1=(\mathrm{u}_{il}^{(1)})\in\mathbb{H}^{n\times n}$. It follows that

$$\sum_{l}u_{il}^{(1)}\sum_{\alpha\in I_{s,n}\{j\}}\left|(\mathbf{AA}^*)_{j.}(\mathbf{e}_{l.})\right|_{\alpha}^{\alpha}=\sum_{\alpha\in I_{s,n}\{j\}}\left|(\mathbf{AA}^*)_{j.}\left(\mathbf{u}_{i.}^{(1)}\right)\right|_{\alpha}^{\alpha},$$

where $\mathrm{u}_{1.}^{(1)}$ is the ith row of $\mathbf{U}_1$. So, we get (18). If we first consider

$$u_{if}^{(2)}:=\sum_{l}\tilde{a}_{fl}\sum_{\alpha\in I_{s,n}\{j\}}\left|(\mathbf{AA}^*)_{j.}(\mathbf{e}_{l.})\right|_{\alpha}^{\alpha}=\sum_{\alpha\in I_{s,n}\{j\}}\left|(\mathbf{AA}^*)_{j.}(\tilde{\mathbf{a}}_{f.})\right|_{\alpha}^{\alpha},\quad f,j=1,\dots,n.$$

and construct the matrix $\mathrm{U}_2=(\mathrm{u}_{if}^{(2)})\in\mathbb{H}^{n\times n}$. then from

$$\sum_{f=1}^{n}\sum_{\beta\in J_{s,n}\{i\}}\left|(\mathbf{A}^2)_{.i}(\mathbf{e}_{.f})\right|_{\beta}^{\beta}u_{if}^{(2)}=\sum_{\beta\in J_{s,n}\{i\}}\left|(\mathbf{A}^2)_{.i}\left(\mathbf{u}_{.f}^{(2)}\right)\right|_{\beta}^{\beta},$$

it follows (19).

Taking into account (17), the following theorem on the determinantal representation of the left core inverse can be proved similarly.

Theorem 3.7. *Let* $\mathrm{A}\in\mathbb{C}_n^{CM}$ *and* rk $\mathbf{A}^2$=rk $\mathbf{A}$=s *Then for its left core inverse* $({}_{\#}\mathrm{A})=(\mathrm{a}_{ij}^{\#,l})$, *we have*

$$a_{ij}^{\#,l}=\frac{\sum_{\alpha\in I_{s,n}\{j\}}\left|(\mathbf{A}^2)_{j.}\left(\mathbf{v}_{i.}^{(1)}\right)\right|_{\alpha}^{\alpha}}{\sum_{\beta\in J_{s,n}}\left|\mathbf{A}^*\mathbf{A}\right|_{\beta}^{\beta}\sum_{\alpha\in I_{s,n}}\left|\mathbf{A}^2\right|_{\alpha}^{\alpha}}=\frac{\sum_{\beta\in J_{s,n}\{i\}}\left|(\mathbf{A}^*\mathbf{A})_{.i}\left(\mathbf{v}_{.j}^{(2)}\right)\right|_{\beta}^{\beta}}{\sum_{\beta\in J_{s,n}}\left|\mathbf{A}^*\mathbf{A}\right|_{\beta}^{\beta}\sum_{\alpha\in I_{s,n}}\left|\mathbf{A}^2\right|_{\alpha}^{\alpha}},$$

where

$$\mathbf{v}_{i.}^{(1)} = \left[\sum_{\beta \in J_{s,n}\{i\}} \left| (\mathbf{A}^* \mathbf{A})_{.i} (\bar{\mathbf{a}}_{.f}) \right|_{\beta}^{\beta} \right] \in \mathbb{C}^{1\times n}, \ f = 1, ..., n$$

$$\mathbf{v}_{.j}^{(2)} = \left[\sum_{\alpha \in I_{s,n}\{j\}} \left| (\mathbf{A}^2)_{j.} (\bar{\mathbf{a}}_{l.}) \right|_{\alpha}^{\alpha} \right] \in \mathbb{C}^{n\times 1}, \ l = 1, ..., n.$$

Here $\bar{\mathbf{a}}_{.f}$ and $\bar{\mathbf{a}}_{l.}$ are the *f*th column and *l*th row of $\bar{\mathbf{A}} := \mathbf{A}^* \mathbf{A}^2$.

3.2 Determinantal representations of the core-EP inverses

Similar as in [4], we introduce two core-EP inverses.

Definition 3.8. A matrix $X \in \mathbb{C}^{n\times n}$ is said to be the right core-EP inverse of $A \in \mathbb{C}^{n\times n}$ if it satisfies the conditions

$$\mathbf{XAX} = \mathbf{A}, and\ \mathcal{C}(\mathbf{X}) = \mathcal{C}(\mathbf{X}^*) = \mathcal{C}\left(\mathbf{A}^d\right).$$

It is denoted as $\mathbf{A}^{\textcircled{\dagger}}$.

Definition 3.9. A matrix $X \in \mathbb{C}^{n\times n}$ is said to be the left core-EP inverse of $A \in \mathbb{C}^{n\times n}$ if it satisfies the conditions

$$\mathbf{XAX} = \mathbf{A}, and\ \mathcal{R}(\mathbf{X}) = \mathcal{R}(\mathbf{X}^*) = \mathcal{R}\left(\mathbf{A}^d\right)$$

It is denoted as $\mathbf{A}_{\textcircled{\dagger}}$.

Remark 3.10. Since $C((A^*)^d) = R(A^d)$, then the left core inverse $\mathbf{A}_{\textcircled{\dagger}}$ of $A \in \mathbb{C}^{n\times n}$ is similar to the ∗core inverse introduced in [4], and the dual core-EP inverse introduced in [35].

Due to [4], we have the following representations the core-EP inverses of $A \in \mathbb{C}^{n\times n}$,

$$\mathbf{A}^{\textcircled{\dagger}} = \mathbf{A}^{\{2,3,6a\}} \ \text{and} \ \ \mathcal{C}(\mathbf{A}^{\textcircled{\dagger}}) \subseteq \mathcal{C}\left(\mathbf{A}^k\right),$$

$$\mathbf{A}_{\textcircled{\dagger}} = \mathbf{A}^{\{2,4,6b\}} \ \text{and} \ \ \mathcal{R}(\mathbf{A}_{\textcircled{\dagger}}) \subseteq \mathcal{R}\left(\mathbf{A}^k\right).$$

Thanks to [35], the following representations of the core-EP inverses will be used for their determinantal representations.

Lemma 3.11. *Let* $A\in\mathbb{C}^{n\times n}$ *and* Ind **A**=*k. Then*

$$\mathbf{A}^{\oplus}=\mathbf{A}^{k}\left(\mathbf{A}^{k+1}\right)^{\dagger}, \tag{21}$$

$$\mathbf{A}_{\oplus}=\left(\mathbf{A}^{k+1}\right)^{\dagger}\mathbf{A}^{k}. \tag{22}$$

Moreover, if Ind **A**=1, then we have the following representations of the right and left core inverses

$$A^{\circledast}=\mathrm{A}(\mathrm{A}^{2})^{\dagger}, \tag{23}$$

$$A_{\circledast}=(\mathrm{A}^{2})^{\dagger}\mathrm{A}. \tag{24}$$

Theorem 3.12. *Suppose* $A\in\mathbb{C}^{n\times n}$, Ind **A**=k, rk$A^k$=s, *and there exist* $\mathbf{A}^{\oplus}$ *and* $\mathbf{A}_{\oplus}$. *Then* $\mathbf{A}^{\oplus}=\left(a_{ij}^{\oplus,r}\right)$ *and* $\mathbf{A}_{\oplus}=\left(a_{ij}^{\oplus,l}\right)$ *possess the determinantal representations, respectively,*

$$a_{ij}^{\oplus,r}=\frac{\sum_{\alpha\in I_{s,n}\{j\}}\left|\left(\mathbf{A}^{k+1}\left(\mathbf{A}^{k+1}\right)^{*}\right)_{j.}(\hat{\mathbf{a}}_{i.})\right|_{\alpha}^{\alpha}}{\sum_{\alpha\in I_{s,n}}\left|\mathbf{A}^{k+1}\left(\mathbf{A}^{k+1}\right)^{*}\right|_{\alpha}^{\alpha}}, \tag{25}$$

$$a_{ij}^{\oplus,l}=\frac{\sum_{\beta\in J_{s,n}\{i\}}\left|\left(\left(\mathbf{A}^{k+1}\right)^{*}\mathbf{A}^{k+1}\right)_{.i}(\check{\mathbf{a}}_{.j})\right|_{\beta}^{\beta}}{\sum_{\beta\in J_{s,n}}\left|\left(\mathbf{A}^{k+1}\right)^{*}\mathbf{A}^{k+1}\right|_{\beta}^{\beta}}, \tag{26}$$

where $\hat{a}_{i.}$ is the ith row of $\hat{\mathrm{A}}=\mathrm{A}^{k}\left(\mathrm{A}^{k+1}\right)^{*}$ and $\check{\mathbf{a}}_{.j}$ is the jth column of $\check{\mathrm{A}}=\left(\mathrm{A}^{k+1}\right)^{*}\mathrm{A}^{k}$

Proof. Consider $\left(\mathbf{A}^{k+1}\right)^{\dagger}=\left(a_{ij}^{(k+1,\dagger)}\right)$ and $\mathbf{A}^{k}=\left(a_{ij}^{(k)}\right)$. By (21),

$$a_{ij}^{\oplus,r}=\sum_{t=1}^{n}a_{it}^{(k)}a_{tj}^{(k+1,\dagger)}.$$

Taking into account (9) for the determinantal representation of $\left(\mathrm{A}^{k+1}\right)^{\dagger}$, we get

$$a_{ij}^{\oplus,r}=\sum_{t=1}^{n}a_{it}^{(k)}\frac{\sum_{\alpha\in I_{s,n}\{j\}}\left|\left(\mathbf{A}^{k+1}\left(\mathbf{A}^{k+1}\right)^{*}\right)_{j.}\left(\mathbf{a}_{t.}^{(k+1,*)}\right)\right|_{\alpha}^{\alpha}}{\sum_{\alpha\in I_{r,m}}\left|\mathbf{A}^{k+1}\left(\mathbf{A}^{k+1}\right)^{*}\right|_{\alpha}^{\alpha}},$$

where $\mathrm{a}_{t.}^{(k+1,*)}$ is the tth row of $\left(\mathrm{A}^{k+1}\right)^{*}$. Since $\sum_{t=1}^{n}a_{it}^{(k)}\mathbf{a}_{t.}^{(k+1,*)}=\hat{\mathbf{a}}_{i.}$, then it follows (25).

The determinantal representation (26) can be obtained similarly by integrating (8) for the determinantal representation of $\left(A^{k+1}\right)^{\dagger}$ in (22).

Taking into account the representations (23)-(24), we obtain the determinantal representations of the right and left core inverses that have more simpler expressions than they are obtained in Theorems 3.6 and 3.7.

Corollary 3.13. $A \in \mathbb{C}_s^{n\times n}$, Ind **A**=1, *and there exist* $A^{\oplus}$ *and* $A_{\oplus}$ *Then* $A^{\oplus}$ $\left(a_{ij}^{\oplus,r}\right)$ *and* $A_{\oplus} = \left(a_{ij}^{\oplus,l}\right)$ *can be expressed as follows*

$$a_{ij}^{\oplus,r} = \frac{\sum_{\alpha \in I_{s,n}\{j\}} \left| \left(\mathbf{A}^2 (\mathbf{A}^2)^*\right)_{j.} (\hat{\mathbf{a}}_{i.}) \right|_{\alpha}^{\alpha}}{\sum_{\alpha \in I_{s,n}} \left| \mathbf{A}^2 (\mathbf{A}^2)^* \right|_{\alpha}^{\alpha}},$$

$$a_{ij}^{\oplus,l} = \frac{\sum_{\beta \in J_{s,n}\{i\}} \left| \left((\mathbf{A}^2)^* \mathbf{A}^2\right)_{.i} (\check{\mathbf{a}}_{.j}) \right|_{\beta}^{\beta}}{\sum_{\beta \in J_{s,n}} \left| (\mathbf{A}^2)^* \mathbf{A}^2 \right|_{\beta}^{\beta}},$$

where $\hat{\mathbf{a}}_{i.}$ is the ith row of $\hat{A} = A(A^2)^*$ and $\check{a}_{.j}$ is the jth column of $\check{A} = (A^2)^*$

3.3 Determinantal representations of the DMP and MPD inverses

The concept of the DMP inverse in complex matrices was introduced in [2] by S. Malik and N. Thome.

Definition 3.14. [2] Suppose $A \in \mathbb{C}^{n\times n}$ and Ind **A** $=k$. A matrix $X \in \mathbb{C}^{n\times n}$ is said to be the DMP inverse of **A** if it satisfies the conditions

$$XAX = X, XA = A^d A, \textit{and } A^k X = A^k A^{\dagger}. \tag{27}$$

It is denoted as $A^{d,\dagger}$.

Due to [2], if an arbitrary matrix satisfies the system of Eq. (27), then it is unique and has the following representation

$$A^{d,\dagger} = A^d A A^{\dagger}. \tag{28}$$

Theorem 3.15. *Let* $A \in \mathbb{C}_s^{n\times n}$, Ind **A**=*k*, *and* rk(A^k)=s_1. *Then, its DMP inverse* $A^{d,\dagger} = \left(a_{ij}^{d,\dagger}\right)$ *has the following determinantal representations.*

$$a_{ij}^{d,\dagger} = \frac{\sum_{\alpha \in I_{s,n}\{j\}} \left| (\mathbf{A}\mathbf{A}^*)_{j.} \left(\mathbf{u}_{i.}^{(1)}\right) \right|_{\alpha}^{\alpha}}{\sum_{\beta \in J_{s_1,n}} \left| \mathbf{A}^{k+1} \right|_{\beta}^{\beta} \sum_{\alpha \in I_{s,n}} \left| \mathbf{A}\mathbf{A}^* \right|_{\alpha}^{\alpha}} = \tag{29}$$

$$= \frac{\sum_{\beta \in J_{s_1,n}\{i\}} \left| \left(\mathbf{A}^{k+1}\right)_{.i} \left(\mathbf{u}_{.j}^{(2)}\right) \right|_{\beta}^{\beta}}{\sum_{\beta \in J_{s_1,n}} \left|\mathbf{A}^{k+1}\right|_{\beta}^{\beta} \sum_{\beta \in J_{s,n}} \left|\mathbf{A}\mathbf{A}^*\right|_{\beta}^{\beta}}, \tag{30}$$

where

$$\mathbf{u}_{i.}^{(1)} = \left[\sum_{\beta \in J_{s_1,n}\{i\}} \left| \left(\mathbf{A}^{k+1}\right)_{.i} (\tilde{\mathbf{a}}_{.f}) \right|_{\beta}^{\beta} \right] \in \mathbb{C}^{1\times n},\ f = 1, ..., n,$$

$$\mathbf{u}_{.j}^{(2)} = \left[\sum_{\alpha \in I_{s,n}\{j\}} \left| (\mathbf{A}\mathbf{A}^*)_{j.} (\tilde{\mathbf{a}}_{l.}) \right|_{\alpha}^{\alpha} \right] \in \mathbb{C}^{n\times 1},\ l = 1, ..., n.$$

Here, $\tilde{\mathbf{a}}_{.f}$ and $\hat{\mathbf{a}}_{l.}$ are the fth column and the lth row of $\tilde{\mathbf{A}} := \mathbf{A}^{k+1}\mathbf{A}^*$.

Proof. Taking into account (28) for $\mathrm{A}^{d,\dagger}$, we get

$$a_{ij}^{d,\dagger} = \sum_{l=1}^{n} \sum_{f=1}^{n} a_{il}^{d} a_{lf} a_{fj}^{\dagger}. \tag{31}$$

By substituting (12) and (9) for the determinantal representations of $\mathbf{A}^d$ and $\mathbf{A}^\dagger$ in (31), we get $a_{ij}^{d,\dagger} =$

$$\sum_{l=1}^{n} \sum_{f=1}^{n} \frac{\sum_{\beta \in J_{s_1,n}\{i\}} \left| \left(\mathbf{A}^{k+1}\right)_{.i} \left(\mathbf{a}_{.l}^{(k)}\right) \right|_{\beta}^{\beta}}{\sum_{\beta \in J_{s_1,n}} \left|\mathbf{A}^{k+1}\right|_{\beta}^{\beta}} a_{lf} \frac{\sum_{\alpha \in I_{s,n}\{j\}} \left| (\mathbf{A}\mathbf{A}^*)_{j.} \left(\mathbf{a}_{f.}^*\right) \right|_{\alpha}^{\alpha}}{\sum_{\alpha \in I_{s,n}} \left|\mathbf{A}\mathbf{A}^*\right|_{\alpha}^{\alpha}} = \tag{32}$$

$$\sum_{l=1}^{n} \sum_{f=1}^{n} \frac{\sum_{\beta \in J_{s_1,n}\{i\}} \left| \left(\mathbf{A}^{k+1}\right)_{.i} (\mathbf{e}_{.l}) \right|_{\beta}^{\beta}}{\sum_{\beta \in J_{s_1,n}} \left|\mathbf{A}^{k+1}\right|_{\beta}^{\beta}} \tilde{a}_{lf} \frac{\sum_{\alpha \in I_{s,n}\{j\}} \left((\mathbf{A}\mathbf{A}^*)_{j.} (\mathbf{e}_{f.}) \right|_{\alpha}^{\alpha}}{\sum_{\alpha \in I_{s,n}} \left|\mathbf{A}\mathbf{A}^*\right|_{\alpha}^{\alpha}},$$

where $\mathrm{e}_{.l}$ and $\mathrm{e}_{l.}$ are the lth unit column and row vectors, and $\tilde{a}_{lf}$ lf is the (lf) th element of the matrix $\tilde{\mathrm{A}} = \mathrm{A}^{k+1}\mathrm{A}^*$. If we put

$$u_{if}^{(1)} := \sum_{l=1}^{n} \sum_{\beta \in J_{s_1,n}\{i\}} \left| \left(\mathbf{A}^{k+1}\right)_{.i} (\mathbf{e}_{.l}) \right|_{\beta}^{\beta} \tilde{a}_{lf} = \sum_{\beta \in J_{s_1,n}\{i\}} \left| \left(\mathbf{A}^{k+1}\right)_{.i} (\tilde{\mathbf{a}}_{.f}) \right|_{\beta}^{\beta},$$

as the fth component of the row vector $\mathbf{u}_{i.}^{(1)} = \left[u_{i1}^{(1)}, ..., u_{in}^{(1)}\right]$, then from

$$\sum_{f=1}^{n} u_{if}^{(1)} \sum_{\alpha \in I_{s,n}\{j\}} \left| (\mathbf{A}\mathbf{A}^*)_{j.} (\mathbf{e}_{f.}) \right|_{\alpha}^{\alpha} = \sum_{\alpha \in I_{s,n}\{j\}} \left| (\mathbf{A}\mathbf{A}^*)_{j.} \left(\mathbf{u}_{i.}^{(1)}\right) \right|_{\alpha}^{\alpha},$$

it follows (29). If we initially obtain

$$u_{lj}^{(2)} := \sum_{f=1}^{n} \bar{a}_{lf} \sum_{\alpha\in I_{s,n}\{j\}} \left|(\mathbf{AA}^*)_{j.}(\mathbf{e}_{f.})\right|_{\alpha}^{\alpha} = \sum_{\alpha\in I_{s,n}\{j\}} \left|(\mathbf{AA}^*)_{j.}(\bar{\mathbf{a}}_{l.})\right|_{\alpha}^{\alpha},$$

as the lth component of the column vector $\mathbf{u}_{.j}^{(2)} = \left[u_{1j}^{(2)}, ..., u_{nj}^{(2)}\right]$, then from

$$\sum_{l=1}^{n} \sum_{\beta\in J_{s_1,n}\{i\}} \left|\left(\mathbf{A}^{k+1}\right)_{.i}(\mathbf{e}_{.l})\right|_{\beta}^{\beta} u_{lj}^{(2)} = \sum_{\beta\in J_{s_1,n}\{i\}} \left|\left(\mathbf{A}^{k+1}\right)_{.i}\left(\mathbf{u}_{.j}^{(2)}\right)\right|_{\beta}^{\beta},$$

it follows (30).

The name of the DMP inverse is in accordance with the order of using the Drazin inverse (D) and the Moore-Penrose (MP) inverse. In that connection, it would be logical to consider the following definition.

Definition 3.16. Suppose $\mathbf{A}\in\mathbb{C}^{n\times n}$ and Ind **A** =k A matrix $\mathbf{X}\in\mathbb{C}^{n\times n}$ is said to be the MPD inverse of **A** if it satisfies the conditions

$$\mathbf{XAX} = \mathbf{X}, \mathbf{AX} = \mathbf{AA}^d, \textit{and } \mathbf{XA}^k = \mathbf{A}^\dagger\mathbf{A}^k.$$

It is denoted as $\mathbf{A}^{\dagger,d}$.

The matrix $\mathbf{A}^{\dagger,d}$ is unique, and it can be represented as

$$\mathbf{A}^{\dagger,d} = \mathbf{A}^\dagger\mathbf{AA}^d. \tag{33}$$

Theorem 3.17. *Let* $\mathbf{A}\in\mathbb{C}^{n\times n}$, Ind **A**=*k*, *and* rk $\mathbf{A}^k$=s_1 *Then, its MPD inverse* $\mathbf{A}^{\dagger,d} = \left(a_{ij}^{\dagger,d}\right)$ *has the following determinantal representations*

$$a_{ij}^{\dagger,d} = \frac{\sum_{\beta\in J_{s,n}\{i\}}\left|(\mathbf{A}^*\mathbf{A})_{.i}\left(\mathbf{v}_{.j}^{(1)}\right)\right|_{\beta}^{\beta}}{\sum_{\beta\in J_{s,n}}|\mathbf{A}^*\mathbf{A}|_{\beta}^{\beta}\sum_{\beta\in I_{s_1,n}}|\mathbf{A}^{k+1}|_{\alpha}^{\alpha}} = \frac{\sum_{\alpha\in I_{s_1,n}\{j\}}\left|\left(\mathbf{A}^{k+1}\right)_{j.}\left(\mathbf{v}_{i.}^{(2)}\right)\right|_{\alpha}^{\alpha}}{\sum_{\alpha\in I_{s_1,n}}|\mathbf{A}^*\mathbf{A}|_{\beta}^{\beta}\sum_{\alpha\in I_{s,n}}|\mathbf{A}^{k+1}|_{\alpha}^{\alpha}},$$

where

$$\mathbf{v}_{.j}^{(1)} = \left[\sum_{\alpha\in I_{s_1,n}\{j\}}\left|\left(\mathbf{A}^{k+1}\right)_{j.}(\hat{\mathbf{a}}_{l.})\right|_{\alpha}^{\alpha}\right] \in\mathbb{C}^{n\times 1},\ l=1,...,n$$

$$\mathbf{v}_{i.}^{(2)} = \left[\sum_{\beta\in J_{s,n}\{i\}}\left|(\mathbf{A}^*\mathbf{A})_{.i}(\hat{\mathbf{a}}_{.f})\right|_{\beta}^{\beta}\right] \in\mathbb{C}^{1\times n},\ l=1,...,n.$$

Here, $\hat{\mathbf{a}}_{l.}$ and $\hat{\mathbf{a}}_{.f}$ are the lth row and the fth column of $\hat{\mathbf{A}} := \mathbf{A}^*\mathbf{A}^{k+1}$.

Proof. The proof is similar to the proof of Theorem 3.15.

3.4 Determinantal representations of the CMP inverse

Definition 3.18. [6] Suppose $A \in \mathbb{C}^{n\times n}$ has the core-nilpotent decomposition $\mathbf{A} = \mathbf{A}_1 + \mathbf{A}_2$, where Ind A_1=Ind A,A_2 is nilpotent, and $A_1A_2=A_2A_1=0$. The CMP inverse of **A** is called the matrix $A^{c,\dagger}:=A^{\dagger}A^{1}A^{\dagger}$.

Lemma 3.19. [6] *Let* $A \in \mathbb{C}^{n\times n}$. *The matrix* $X=A^{c,\dagger}$,†*is the unique matrix that satisfies the following system of equations:*

$$\mathbf{XAX} = \mathbf{X}, \mathbf{AXA} = \mathbf{A}_1, \mathbf{AX} = \mathbf{A}_1\mathbf{A}^{\dagger}, and\ \mathbf{XA} = \mathbf{A}^{\dagger}\mathbf{A}_1.$$

Moreover,

$$\mathbf{A}^{c,\dagger} = \mathbf{Q}_A\mathbf{A}^d\mathbf{P}_A. \tag{34}$$

Taking into account (34), it follows the next theorem about determinantal representations of the quaternion CMP inverse.

Theorem 3.20. *Let* $A \in \mathbb{C}_s^{n\times n}$, Ind A=m, *and* $\mathrm{rk}(A^m)=s_1$. *Then, the determinantal representations of its CMP inverse* $\mathbf{A}^{c,\dagger} = \left(a_{ij}^{c,\dagger}\right)$ *can be expressed as*

$$a_{ij}^{c,\dagger} = \frac{\sum_{\beta \in J_{s,n}\{i\}} \left|(\mathbf{A}^*\mathbf{A})_{.i}\left(\mathbf{v}_{.j}^{(l)}\right)\right|_\beta^\beta}{\left(\sum_{\beta \in J_{s,n}} |\mathbf{A}^*\mathbf{A}|_\beta^\beta\right)^2 \sum_{\beta \in J_{s_1,n}} |\mathbf{A}^{m+1}|_\beta^\beta} \tag{35}$$

$$a_{ij}^{c,\dagger} = \frac{\sum_{\alpha \in I_{s,n}\{j\}} \left|(\mathbf{A}\mathbf{A}^*)_{j.}\left(\mathbf{w}_{i.}^{(l)}\right)\right|_\alpha^\alpha}{\left(\sum_{\alpha \in I_{s,n}} |\mathbf{A}\mathbf{A}^*|_\alpha^\alpha\right)^2 \sum_{\beta \in J_{s_1,n}} |\mathbf{A}^{m+1}|_\beta^\beta} \tag{36}$$

for all l=1,2, where

$$\mathbf{v}_{.j}^{(1)} = \left[\sum_{\alpha \in I_{s,n}\{j\}} \left|(\mathbf{A}\mathbf{A}^*)_{j.}(\hat{\mathbf{u}}_{t.})\right|_\alpha^\alpha\right] \in \mathbb{C}^{n\times 1}, t = 1, \ldots, n, \tag{37}$$

$$\mathbf{w}_{i.}^{(1)} = \left[\sum_{\beta \in J_{s,n}\{i\}} \left|(\mathbf{A}^*\mathbf{A})_{.i}(\hat{\mathbf{u}}_{.k})\right|_\beta^\beta\right] \in \mathbb{C}^{1\times n}, k = 1, \ldots, n, \tag{38}$$

$$\mathbf{v}_{.j}^{(2)} = \left[\sum_{\alpha \in I_{s,n}\{j\}} \left|(\mathbf{A}^*\mathbf{A})_{j.}(\tilde{\mathbf{g}}_{t.})\right|_\alpha^\alpha\right] \in \mathbb{C}^{n\times 1}, t = 1, \ldots, n, \tag{39}$$

$$\mathbf{w}_{i.}^{(2)} = \left[\sum_{\beta \in J_{s,n}\{i\}} \left|(\mathbf{A}^*\mathbf{A})_{.i}(\tilde{\mathbf{g}}_{.k})\right|_\beta^\beta\right] \in \mathbb{C}^{1\times n}, k = 1, \ldots, n. \tag{40}$$

Here, $\hat{\mathbf{u}}_{t.}$ is the tth row and $\hat{\mathbf{u}}_{.k}$ is the kth column of $\hat{\mathrm{U}} := \mathrm{UAA}^*$, $\tilde{\mathbf{g}}_{t.}$ is the tth row and $\tilde{\mathbf{g}}_{.k}$ is the tth column of $\tilde{\mathrm{G}} := \mathbf{A}^*\mathbf{AG}$, and the matrices $\mathrm{U} = (u_{ij}) \in \mathbb{H}^{n\times n}$ and $(g_{ij}) \in \mathbb{H}^{n\times n}$ are such that

$$u_{ij} = \sum_{\alpha\in I_{s_1,n}\{j\}} \left|(\mathbf{A}^{m+1})_{j.}(\hat{\mathbf{a}}_{i.})\right|_{\alpha}^{\alpha}, \quad g_{ij} = \sum_{\beta\in J_{s_1,n}\{i\}} \left|(\mathbf{A}^{m+1})_{.i}(\tilde{\mathbf{a}}_{.j})\right|_{\beta}^{\beta},$$

where $\hat{\mathbf{a}}_{i.}$ is the ith row of $\hat{\mathrm{A}} := \mathbf{A}^*\mathbf{A}^{m+1}$ and $\tilde{\mathbf{a}}_{.j}$ is the jth column of $\tilde{\mathrm{A}} := \mathbf{A}^{m+1}\mathbf{A}^*$.

Proof. Taking into account (34), we get

$$a_{ij}^{c,\dagger} = \sum_{l=1}^{n}\sum_{k=1}^{n} q_{il}^{A} a_{lk}^{d} p_{kj}^{A}, \tag{41}$$

where $\mathrm{Q}_A = (q_{il}^A)$, $\mathbf{A}^d = (a_{il}^d)$, and $\mathbf{P}_A = (p_{il}^A)$.

(a) Taking into account the expressions (13), (10), and (11) for the determinantal representations of A^d, Q^A, and P_A, respectively, we have

$$a_{ij}^{c,\dagger} = \sum_{l}\sum_{t} \frac{\sum_{\beta\in J_{s,n}\{i\}}\left|(\mathbf{A}^*\mathbf{A})_{.i}(\dot{\mathbf{a}}_{.t})\right|_{\beta}^{\beta}}{\sum_{\beta\in J_{s,n}}\left|\mathbf{A}^*\mathbf{A}\right|_{\beta}^{\beta}} \frac{\sum_{\alpha\in I_{s_1,n}\{l\}}\left|(\mathbf{A}^{m-1})_{l.}\left(\mathbf{a}_{t.}^{(m)}\right)\right|_{\alpha}^{\alpha}}{\sum_{\alpha\in I_{s_1,n}}\left|\mathbf{A}^{m-1}\right|_{\alpha}^{\alpha}} \frac{\sum_{\alpha\in I_{s,n}\{j\}}\left|(\mathbf{AA}^*)_{j.}(\ddot{\mathbf{a}}_{l.})\right|_{\alpha}^{\alpha}}{\sum_{\alpha\in I_{s,n}}\left|\mathbf{AA}^*\right|_{\alpha}^{\alpha}},$$

where $\dot{\mathbf{a}}_{.t}$ is the tth column of $\mathbf{A}^*\mathbf{A}$, $\ddot{\mathbf{a}}_{l.}$ is the lth row of AA^*, and $\mathrm{a}_{t.}^{(m)}$ is the tth row of A^m. So, it is clear that

$$a_{ij}^{c,\dagger} = \sum_{l}\sum_{t}\sum_{k} \frac{\sum_{\beta\in J_{s,n}\{i\}}\left|(\mathbf{A}^*\mathbf{A})_{.i}(\mathbf{e}_{.t})\right|_{\beta}^{\beta} \hat{a}_{tk} \sum_{\alpha\in I_{s_1,n}\{l\}}\left|(\mathbf{A}^{m+1})_{l.}(\mathbf{e}_{k.})\right|_{\alpha}^{\alpha}}{\sum_{\beta\in J_{s,n}}\left|\mathbf{A}^*\mathbf{A}\right|_{\beta}^{\beta}\sum_{\alpha\in I_{s_1,n}}\left|\mathbf{A}^{m+1}\right|_{\alpha}^{\alpha}} \frac{\sum_{\alpha\in I_{s,n}\{j\}}\left|(\mathbf{AA}^*)_{j.}(\ddot{\mathbf{a}}_{l.})\right|_{\alpha}^{\alpha}}{\sum_{\alpha\in I_{s,n}}\left|\mathbf{AA}^*\right|_{\alpha}^{\alpha}},$$

where $\mathrm{e}_{.t}$ is the tth unit column vector, $\mathrm{e}_{k.}$ is the kth row vector, and $\hat{\mathrm{a}}_{tk}$ is the (tk)th element of $\hat{\mathrm{A}} = \mathrm{A}^*\mathrm{A}^{m+1}$.

Denote

$$u_{tl} := \sum_{k}\hat{a}_{tk} \sum_{\alpha\in I_{s_1,n}\{j\}}\left|(\mathbf{A}^{m+1})_{l.}(\mathbf{e}_{k.})\right|_{\alpha}^{\alpha} = \sum_{\alpha\in I_{s_1,n}\{j\}}\left|(\mathbf{A}^{m+1})_{l.}(\hat{\mathbf{a}}_{t.})\right|_{\alpha}^{\alpha} \tag{42}$$

as the tth component of a column vector $\mathrm{u}_{.l} = [u_{1l},\ldots,u_{nl}]$ Then from

$$\sum_{t}\sum_{\beta\in J_{s,n}\{i\}}\left|(\mathbf{A}^*\mathbf{A})_{.i}(\mathbf{e}_{.t})\right|_{\beta}^{\beta} u_{tl} = \sum_{\beta\in J_{s,n}\{i\}}\left|(\mathbf{A}^*\mathbf{A})_{.i}(\mathbf{u}_{.l})\right|_{\beta}^{\beta},$$

we have

$$a_{ij}^{c,\dagger} = \sum_{l}\frac{\sum_{\beta\in J_{s,n}\{i\}}\left|(\mathbf{A}^*\mathbf{A})_{.i}(\mathbf{u}_{.l})\right|_{\beta}^{\beta}\sum_{\alpha\in I_{s,n}\{j\}}\left|(\mathbf{AA}^*)_{j.}(\ddot{\mathbf{a}}_{l.})\right|_{\alpha}^{\alpha}}{\sum_{\beta\in J_{s,n}}\left|\mathbf{A}^*\mathbf{A}\right|_{\beta}^{\beta}\sum_{\alpha\in I_{s_1,n}}\left|\mathbf{A}^{m+1}\right|_{\alpha}^{\alpha}\sum_{\alpha\in I_{s,n}}\left|\mathbf{AA}^*\right|_{\alpha}^{\alpha}}.$$

Construct the matrix $\mathrm{U}=(u_{tl}) \in \mathbb{H}^{n\times n}$, where u_{tl} is given by (42), and denote $\hat{\mathrm{U}} := \mathrm{UAA}^*$. Then, taking into account that $\left|\mathbf{A}^*\mathbf{A}\right|_{\beta}^{\beta} = \left|\mathbf{AA}^*\right|_{\alpha}^{\alpha}$, we have

$$a_{ij}^{c,\dagger}=\frac{\sum_t\sum_k\sum_{\beta\in J_{s,n}\{i\}}\left|(\mathrm{A}^*\mathrm{A})_{.i}(\mathrm{e}_{.t})\right|_\beta^\beta \hat{u}_{tk}\sum_{\alpha\in I_{s,n}\{j\}}\left|(\mathrm{AA}^*)_{j.}(\mathrm{e}_{k.})\right|_\alpha^\alpha}{\left(\sum_{\beta\in J_{s,n}}\left|\mathrm{A}^*\mathrm{A}\right|_\beta^\beta\right)^2\sum_{\alpha\in I_{s_1,n}}\left|\mathrm{A}^{m+1}\right|_\alpha^\alpha}$$

If we put that

$$v_{tj}^{(1)}:=\sum_k \hat{u}_{tk}\sum_{\alpha\in I_{s,n}\{j\}}\left|(\mathrm{AA}^*)_{j.}(\mathrm{e}_{k.})\right|_\alpha^\alpha=\sum_{\alpha\in I_{s,n}\{j\}}\left|(\mathrm{AA}^*)_{j.}(\hat{\mathrm{u}}_{t.})\right|_\alpha^\alpha$$

is the tth component of a column vector $\mathrm{v}_{.j}^{(1)}=\left[v_{1j}^{(1)},\dots,v_{nj}^{(1)}\right]$, then from

$$\sum_t\sum_{\beta\in J_{s,n}\{i\}}\left|(\mathrm{A}^*\mathrm{A})_{.i}(\mathrm{e}_{.t})\right|_\beta^\beta v_{tj}^{(1)}=\sum_{\beta\in J_{s,n}\{i\}}\left|(\mathrm{A}^*\mathrm{A})_{.i}\left(\mathrm{v}_{.j}^{(1)}\right)\right|_\beta^\beta,$$

it follows (35) with $\mathrm{v}_{.j}^{(1)}$ given by (37). If we initially put

$$w_{ik}^{(1)}:=\sum_t\sum_{\beta\in J_{s,n}\{i\}}\left|(\mathrm{A}^*\mathrm{A})_{.i}(\mathrm{e}_{.t})\right|_\beta^\beta \hat{u}_{tk}=\sum_{\beta\in J_{s,n}\{i\}}\left|(\mathrm{A}^*\mathrm{A})_{.i}(\hat{\mathrm{u}}_{.k})\right|_\beta^\beta$$

as the kth component of the row vector $\mathrm{w}_{i.}^{(1)}=\left[w_{i1}^{(1)},\dots,w_{in}^{(1)}\right]$, then from

$$\sum_k w_{ik}^{(1)}\sum_{\alpha\in I_{s,n}\{j\}}\left|(\mathrm{A}^2)_{j.}(\mathrm{e}_{k.})\right|_\alpha^\alpha=\sum_{\alpha\in I_{s,n}\{j\}}\left|(\mathrm{A}^2)_{j.}\left(\mathrm{w}_{i.}^{(1)}\right)\right|_\alpha^\alpha,$$

it follows (36) with $\mathrm{w}_{i.}^{(1)}$ given by (38).

(b) By using the determinantal representation (12) for A^{d} in (41), we have

$$a_{ij}^{c,\dagger}=\sum_k\sum_t\frac{\sum_{\beta\in J_{s,n}\{i\}}\left|(\mathrm{A}^*\mathrm{A})_{.i}(\dot{\mathrm{a}}_{.t})\right|_\beta^\beta}{\sum_{\beta\in J_{s,n}}\left|\mathrm{A}^*\mathrm{A}\right|_\beta^\beta}\frac{\sum_{\beta\in J_{s_1,n}\{t\}}\left|(\mathrm{A}^{m+1})_{.t}\left(\mathrm{a}_{.k}^{(m)}\right)\right|_\beta^\beta}{\sum_{\beta\in J_{s_1,n}}\left|\mathrm{A}^{m+1}\right|_\beta^\beta}\frac{\sum_{\alpha\in I_{s,n}\{j\}}\left|(\mathrm{AA}^*)_{j.}(\ddot{\mathrm{a}}_{k.})\right|_\alpha^\alpha}{\sum_{\alpha\in I_{s,n}}\left|\mathrm{AA}^*\right|_\alpha^\alpha}.$$

Therefore,

$$a_{ij}^{c,\dagger}=\sum_l\sum_k\sum_t\frac{\sum_{\beta\in J_{s,n}\{i\}}\left((\mathrm{A}^*\mathrm{A})_{.i}(\dot{\mathrm{a}}_{.t})\right)_\beta^\beta}{\sum_{\beta\in J_{s,n}}\left|\mathrm{A}^*\mathrm{A}\right|_\beta^\beta}\times$$

$$\frac{\sum_{\beta\in J_{s_1,n}\{t\}}\left|(\mathrm{A}^{m+1})_{.t}(\mathrm{e}_{.k})\right|_\beta^\beta}{\sum_{\beta\in J_{s_1,n}}\left|\mathrm{A}^{m+1}\right|_\beta^\beta}\tilde{a}_{kl}\frac{\sum_{\alpha\in I_{s,n}\{j\}}\left|(\mathrm{AA}^*)_{j.}(\mathrm{e}_{l.})\right|_\alpha^\alpha}{\sum_{\alpha\in I_{s,n}}\left|\mathrm{AA}^*\right|_\alpha^\alpha}.$$

where $\mathrm{e}_{.k}$ is the kth unit column vector, $\mathrm{e}_{l.}$ is the lth unit row vector, and $\tilde{a}_{kl}$ is the (kl)th element of $\tilde{\mathrm{A}}=\mathrm{A}^{m+1}\mathrm{A}^*$.

If we denote

$$g_{tl}:=\sum_l\sum_{\beta\in J_{s_1,n}\{t\}}\left|(\mathrm{A}^{m+1})_{.t}(\mathrm{e}_{.k})\right|_\beta^\beta\tilde{a}_{kl}=\sum_{\beta\in J_{s_1,n}\{t\}}\left|(\mathrm{A}^{m+1})_{.t}(\tilde{\mathrm{a}}_{.l})\right|_\beta^\beta \qquad (43)$$

as the lth component of a row vector $\mathbf{g}_{t.} = [g_{t1}, ..., g_{tm}]$, then

$$\sum_l g_{tl} \sum_{\alpha \in I_{s,\kappa}\{j\}} \left|(\mathbf{AA}^*)_{j.}(\mathbf{e}_{l.})\right|_\alpha^\alpha = \sum_{\alpha \in I_{s,\kappa}\{j\}} \left|(\mathbf{AA}^*)_{j.}(\mathbf{g}_{t.})\right|_\alpha^\alpha.$$

From this, it follows that

$$a_{ij}^{c,\dagger} = \sum_t \frac{\sum_{\beta \in J_{s,\kappa}\{i\}} |(\mathbf{A}^*\mathbf{A})_{.i}(\dot{\mathbf{a}}_{.t})|_\beta^\beta \sum_{\alpha \in I_{s,\kappa}\{j\}} |(\mathbf{AA}^*|_{j.}(\mathbf{g}_{t.}))_\alpha^\alpha}{\sum_{\beta \in J_{r,\kappa}} |\mathbf{A}^*\mathbf{A}|_\beta^\beta \sum_{\alpha \in I_{r_1,\kappa}} |\mathbf{A}^{m+1}|_\alpha^\alpha \sum_{\alpha \in I_{s,\kappa}} |\mathbf{AA}^*|_\alpha^\alpha}$$

Construct the matrix $G=(g_{tl})\in\mathbb{H}^{n\times n}$, where g_{tl} is given by (43). Denote $\tilde{\mathrm{G}} := \mathrm{A}^*\mathrm{AG}$. Then,

$$a_{ij}^{c,\dagger} = \frac{\sum_t \sum_k \sum_{\beta \in J_{s,\kappa}\{i\}} |(\mathbf{A}^*\mathbf{A})_{.i}(\mathbf{e}_{.t})|_\beta^\beta \tilde{g}_{tk} \sum_{\alpha \in I_{s,\kappa}\{j\}} \left|(\mathbf{AA}^*)_{j.}(\mathbf{e}_{k.})\right|_\alpha^\alpha}{\sum_{\beta \in J_{s,\kappa}} \left(|\mathbf{A}^*\mathbf{A}|_\beta^\beta\right)^2 \sum_{\alpha \in I_{r_1,\kappa}} |\mathbf{A}^{m+1}|_\alpha^\alpha}$$

If we denote

$$v_{tj}^{(2)} := \sum_k \tilde{g}_{tk} \sum_{\alpha \in I_{s,\kappa}\{j\}} \left|(\mathbf{AA}^*)_{j.}(\mathbf{e}_{k.})\right|_\alpha^\alpha = \sum_{\alpha \in I_{s,\kappa}\{j\}} \left|(\mathbf{AA}^*)_{j.}(\tilde{\mathbf{g}}_{t.})\right|_\alpha^\alpha$$

as the tth component of a column vector $\mathbf{v}_{.j}^{(2)} = \left[v_{1j}^{(2)}, ..., v_{nj}^{(2)}\right]$, then

$$\sum_t \sum_{\beta \in J_{s,\kappa}\{i\}} |(\mathbf{A}^*\mathbf{A})_{.i}(\mathbf{e}_{.t})|_\beta^\beta v_{tj}^{(2)} = \sum_{\beta \in J_{s,\kappa}\{i\}} \left|(\mathbf{A}^*\mathbf{A})_{.i}\left(\mathbf{v}_{.j}^{(2)}\right)\right|_\beta^\beta.$$

Thus, we have (35) with $\mathbf{v}_{.j}^{(2)}$ given by (39).

If, now, we denote

$$w_{ik}^{(2)} := \sum_t \sum_{\beta \in J_{s,\kappa}\{i\}} |(\mathbf{A}^*\mathbf{A})_{.i}(\mathbf{e}_{.t})|_\beta^\beta \tilde{g}_{tk} = \sum_{\beta \in J_{s,\kappa}\{i\}} |(\mathbf{A}^*\mathbf{A})_{.i}(\tilde{\mathbf{g}}_{.k})|_\beta^\beta$$

as the kth component of a row vector $\mathbf{w}_{i.}^{(2)} = \left[w_{i1}^{(2)}, ..., w_{in}^{(2)}\right]$ then

$$\sum_k w_{ik}^{(2)} \sum_{\alpha \in I_{s,\kappa}\{j\}} \left|(\mathbf{AA}^*)_{j.}(\mathbf{e}_{k.})\right|_\alpha^\alpha = \sum_{\alpha \in I_{s,\kappa}\{j\}} \left|(\mathbf{AA}^*)_{j.}\left(\mathbf{w}_{i.}^{(2)}\right)\right|_\alpha^\alpha.$$

So, finally, we have (36) with $\mathbf{w}_{i.}^{(2)}$ given by (40).

4. AN EXAMPLE

Given the matrix

$$\mathrm{A} = \begin{bmatrix} 2 & 0 & 0 \\ -i & i & i \\ -i & -i & -i \end{bmatrix}.$$

Since

$$AA^* = \begin{bmatrix} 4 & 2i & 2i \\ -2i & 3 & -1 \\ -2i & -1 & 3 \end{bmatrix}, \ A^2 = \begin{bmatrix} 4 & 0 & 0 \\ 2-2i & 0 & 0 \\ -2-2i & 0 & 0 \end{bmatrix}, \ A^3 = \begin{bmatrix} 8 & 0 & 0 \\ 4-4i & 0 & 0 \\ -4-4i & 0 & 0 \end{bmatrix};$$

then rkA=2 and rkA^2=rk A^3=1, and k=Ind A=2 and r_1=1. So, we shall find $A^{\oplus}$ and $A_{\oplus}$ by (25) and (26), respectively.

Since

$$\hat{A} = A^2(A^3)^* = 16\begin{bmatrix} 2 & 1+i & -1+i \\ 1-i & 1 & i \\ -1-i & i & 1 \end{bmatrix}$$

then by (25),

$$a_{11}^{\oplus,r} = \frac{\sum_{\alpha\in I_{1,3}\{1\}}\left|\left(A^3(A^3)^*\right)_{1.}(\hat{a}_{1.})\right|_{\alpha}^{\alpha}}{\sum_{\alpha\in I_{1,3}}\left|A^3(A^3)^*\right|_{\alpha}^{\alpha}} = \frac{1}{4}.$$

By similarly continuing, we get

$$A^{\oplus} = \frac{1}{8}\begin{bmatrix} 2 & 1+i & -1+i \\ 1-i & 1 & i \\ -1-i & i & 1 \end{bmatrix}$$

By analogy, due to (26), we have

$$A_{\oplus} = \frac{1}{2}\begin{bmatrix} 1 & 0 & 0 \\ 0 & 0 & 0 \\ 0 & 0 & 0 \end{bmatrix}$$

The DMP inverse $A^{d,\dagger}$ can be found by Theorem 3.15. Since

$$\bar{A} = A^3A^* = 4\begin{bmatrix} 4 & 2i & 2i \\ 2-2i & 1+i & 1+i \\ -2-2i & 1-i & 1-i \end{bmatrix}$$

and rk (A^3)=1, then

$$u_1^{(1)} = \bar{a}_{1.}, \ u_2^{(1)} = \bar{a}_{2.}, \ u_3^{(1)} = \bar{a}_{3.}.$$

Furthermore, by (29),

$$a_{11}^{d,\dagger} = \frac{\sum_{\alpha\in I_{2,3}\{1\}}\left|(AA^*)_{1.}\left(u_{1.}^{(1)}\right)\right|_{\alpha}^{\alpha}}{\sum_{\beta\in J_{1,3}}\left|A^3\right|_{\beta}^{\beta}\sum_{\alpha\in I_{2,3}}\left|AA^*\right|_{\alpha}^{\alpha}} = \frac{1}{192}\left(det\begin{bmatrix} 16 & 8i \\ -2i & 3 \end{bmatrix} + det\begin{bmatrix} 16 & 8i \\ -2i & 3 \end{bmatrix}\right) = \frac{1}{3}.$$

By similarly continuing, we get

$$A^{d,\dagger} = \frac{1}{12}\begin{bmatrix} 4 & 2i & 2i \\ 2-2i & 1+i & 1+i \\ -2-2i & 1-i & 1-i \end{bmatrix}$$

Similarly, by Theorem 3.17, we get

$$A^{\dagger,d} = \frac{1}{4}\begin{bmatrix} 2 & 0 & 0 \\ -i & 0 & 0 \\ -i & 0 & 0 \end{bmatrix}$$

Finally, by theorem, we find the CMP inverse $A^{c,\dagger} = \left(a_{ij}^{c,\dagger}\right)$ Since rk A^3=1, then $G = \tilde{A}$ and

$$\tilde{G} = A^* A\tilde{A} = 16\begin{bmatrix} 6 & 3i & 3i \\ -2i & 1 & 1 \\ -2i & 1 & 1 \end{bmatrix}$$

Furthermore, by (40),

$$w_{11}^{(2)} = \sum_{\beta \in J_{2,3}\{1\}} \left|(A^*A)_{.1}(\tilde{g}_{.1})\right|_{\beta}^{\beta} = \left(det\begin{bmatrix} 6 & 0 \\ -2i & 2 \end{bmatrix} + det\begin{bmatrix} 6 & 0 \\ -2i & 2 \end{bmatrix}\right) = 24.$$

By similar calculations, we get

$$\mathbf{w}_{1.}^{(2)} = [384, 96\mathbf{i}, 96\mathbf{i}],\ \ \mathbf{w}_{2.}^{(2)} = [-192\mathbf{i}, 96, 96],\ \ \mathbf{w}_{3.}^{(2)} = [-192\mathbf{i}, 96, 06].$$

So, by (36), we get

$$a_{11}^{c,\dagger} = \frac{\sum_{\alpha \in I_{2,3}\{1\}} \left|(AA^*)_{1.}\left(\mathbf{w}_{1.}^{(2)}\right)\right|_{\alpha}^{\alpha}}{\left(\sum_{\alpha \in I_{2,3}} |AA^*|_{\alpha}^{\alpha}\right)^2 \sum_{\beta \in J_{1,3}} |A^3|_{\beta}^{\beta}}$$
$$= \frac{1}{4608}\left(det\begin{bmatrix} 384 & 192i \\ -2i & 3 \end{bmatrix} + det\begin{bmatrix} 384 & 192i \\ -2i & 3 \end{bmatrix}\right) = \frac{1}{3}.$$

By similarly continuing, we derive

$$A^{c,\dagger} = \frac{1}{12}\begin{bmatrix} 4 & 2i & 2i \\ -2i & 1 & 1 \\ -2i & 1 & 1 \end{bmatrix}$$

5. CONCISE SUMMARY

This chapter presents a straightforward approach for determining the core inverse and its generalizations, which are derived from their determinantal representations. The right and left core inverses, the right and left core-EP inverses, the DMP, MPD, and CMP inverses are obtained by new determinantal representations.

6. AT A GLANCE

Generalized inverse matrices are significant entities in the field of matrix theory. Specifically, they are valuable instruments in resolving matrix equations. Two well-known generalized inverses are the Moore-Penrose inverse and the Drazin inverse. A novel generalized inverse matrix called the core inverse has been recently introduced. This was then extended to include the core-EP inverse, as well as the BT, DMP, and CMP inverses. Unlike the inverse matrix, which has a determinantal representation based on cofactors, basic generalized inverses can have several determinantal representations due to the search for more appropriate explicit expressions. This chapter presents novel and unique determinantal representations of the core inverse and its generalizations. These representations are derived by utilizing determinantal representations of the Moore-Penrose and Drazin inverses, which were previously found by the author.

REFERENCES

1. Baksalary OM, Trenkler G. Core inverse of matrices. Linear and Multilinear Algebra. 2010;58:681-697
2. Malik S, Thome N. On a new generalized inverse for matrices of an arbitrary index. Applied Mathematics and Computation. 2014;226:575-580
3. Xu SZ, Chen JL, Zhang XX. New characterizations for core inverses in rings with involution. Frontiers in Mathematics China. 2017;12:231-246
4. Prasad KM, Mohana KS. Core EP inverse. Linear and Multilinear Algebra. 2014;62(3):792-802
5. Baksalary OM, Trenkler G. On a generalized core inverse. Applied Mathematics and Computation. 2014;236:450-457
6. Mehdipour M, Salemi A. On a new generalized inverse of matrices. Linear and Multilinear Algebra. 2018;66(5):1046-1053
7. Chen JL, Zhu HH, Patrićio P, Zhang YL. Characterizations and representations of core and dual core inverses. Canadian Mathematical Bulletin. 2017;60:269-282
8. Gao YF, Chen JL. Pseudo core inverses in rings with involution. Communications in Algebra. 2018;46:38-50
9. Guterman A, Herrero A, Thome N. New matrix partial order based on spectrally orthogonal matrix decomposition. Linear and Multilinear Algebra. 2016;64(3):362-374
10. Ferreyra DE, Levis FE, Thome N. Maximal classes of matrices determining generalized inverses. Applied Mathematics and Computation. 2018;333:42-52

11. Ferreyra DE, Levis FE, Thome N. Revisiting the core EP inverse and its extension to rectangular matrices. Quaestiones Mathematicae. 2018;41(2):265-281
12. Liu X, Cai N. High-order iterative methods for the DMP inverse. Journal of Mathematics. 2018;8175935:6
13. Ma H, Stanimirović PS. Characterizations, approximation and perturbations of the core-EP inverse. Applied Mathematics and Computation. 2019;359:404-417
14. Mielniczuk J. Note on the core matrix partial ordering. Discussiones Mathematicae Probability and Statistics 2011;31:71-75
15. Mosić D, Deng C, Ma H. On a weighted core inverse in a ring with involution. Communications in Algebra. 2018;46(6):2332-2345
16. Prasad KM, Raj MD. Bordering method to compute core-EP inverse. Special Matrices. 2018;6:193-200
17. Rakić DS, Dinčić ČN, Djordjević DS. Group, Moore-Penrose, core and dual core inverse in rings with involution. Linear Algebra and its Applications. 2014;463:115-133
18. Wang HX. Core-EP decomposition and its applications. Linear Algebra and its Applications. 2016;508:289-300
19. Bapat RB, Bhaskara Rao KPS, Prasad KM. Generalized inverses over integral domains. Linear Algebra and its Applications. 1990;140:181-196
20. Bhaskara Rao KPS. Generalized inverses of matrices over integral domains. Linear Algebra and its Applications. 1983;49:179-189
21. Kyrchei I. Analogs of the adjoint matrix for generalized inverses and corresponding Cramer rules. Linear and Multilinear Algebra. 2008;56(4):453-469
22. Kyrchei I. Explicit formulas for determinantal representations of the Drazin inverse solutions of some matrix and differential matrix equations. Applied Mathematics and Computation. 2013;219:7632-7644
23. Kyrchei I. Cramer's rule for generalized inverse solutions. In: Kyrchei I, editor. Advances in Linear Algebra Research. New York: Nova Science Publ; 2015. pp. 79-132
24. Stanimirović PS. General determinantal representation of pseudoinverses of matrices. Matematichki Vesnik. 1996;48:1-9
25. Stanimirović PS, Djordjevic DS. Full-rank and determinantal representation of the Drazin inverse. Linear Algebra and its Applications. 2000;311:131-151

26. Kyrchei I. Determinantal representations of the Moore-Penrose inverse over the quaternion skew field. Journal of Mathematical Sciences. 2012;180(1):23-33

27. Kyrchei I. Determinantal representations of the Moore-Penrose inverse over the quaternion skew field and corresponding Cramer's rules. Linear and Multilinear Algebra. 2011;59(4):413-431

28. Kyrchei I. Determinantal representations of the Drazin inverse over the quaternion skew field with applications to some matrix equations. Applied Mathematics and Computation. 2014;238:193-207

29. Kyrchei I. Determinantal representations of the W-weighted Drazin inverse over the quaternion skew field. Applied Mathematics and Computation. 2015;264:453-465

30. Kyrchei I. Explicit determinantal representation formulas of W-weighted Drazin inverse solutions of some matrix equations over the quaternion skew field. Mathematical Problems in Engineering. 2016;8673809:13

31. Kyrchei I. Explicit determinantal representation formulas for the solution of the two-sided restricted quaternionic matrix equation. Journal of Applied Mathematics and Computing. 2018;58(1-2):335-365

32. Kyrchei I. Determinantal representations of the Drazin and W-weighted Drazin inverses over the quaternion skew field with applications. In: Griffin S, editor. Quaternions: Theory and Applications. New York: Nova Sci. Publ.; 2017. pp. 201-275

33. Kyrchei I. Weighted singular value decomposition and determinantal representations of the quaternion weighted Moore-Penrose inverse. Applied Mathematics and Computation. 2017;309:1-16

34. Kyrchei I. Determinantal representations of the quaternion weighted Moore-Penrose inverse and its applications. In: Baswell AR, editor. Advances in Mathematics Research 23. New York: Nova Science Publ; 2017. pp. 35-96

35. Zhou M, Chen J, Li T, Wang D. Three limit representations of the core-EP inverse. Univerzitet u Nišu. 2018;32:5887-5894

02 MATRIX EXPONENTIALS AND LINEAR SYSTEMS OF ALGEBRAIC EQUATIONS: NEW FORMULAS

1. INTRODUCTION: AN OVERVIEW

The recently developed new matrix series expressions [1] for solving basic first order differential equations that are related with general systems of linear algebraic equations. The above differential equations reflect the perpendicular paths of a group of hypersurfaces that depict a quadratic functional associated with the linear algebraic system. To find the solution of the latter, one can reduce the functional along an orthogonal trajectory instead of using various strategies that involve minimizing along conjugate gradient directions or using minimized iterations [2]. Given that the solutions of the differential equations in question are directly connected to the solutions of the corresponding algebraic systems through matrix exponentials, it is possible to develop efficient solution methods if accurate and computationally efficient numerical calculations using matrix exponentials can be employed. The paper [3] provides a comprehensive assessment of different algorithms used to calculate matrix exponentials, along with a detailed discussion of the challenges associated with this task.

This chapter employs innovative formulas for arbitrary matrix exponentials that integrate strongly convergent infinite series, allowing for accurate and reliable numerical calculations. By employing these mathematical equations, two innovative solution methods are developed that demonstrate outstanding efficiency when used on large-scale generic linear algebraic systems.

2. DIFFERENTIAL EQUATIONS ASSOCIATED WITH LINEAR SYSTEMS OF ALGEBRAICEQUATIONS

We start with simple vector differential equations whose solutions are related to the solution of general systems of linear algebraic equations. Later, we construct differential equations that allow an efficient numerical integration in order to obtain the solution of these systems.

Matrix series solution of some vector differential equations

Consider a first order vector differential equation of the form

$$\frac{dx}{dv} = f(v)(Ax - b) \tag{1}$$

where $A \in R^{n\times n}$ is a general nonsingular matrix, $b \in R^n$ is a given n-dimensional vector, $x=(x_1,x_2,...x_n)^T$ is the unknown n-dimensional vector, T indicates the transpose, and $f(v)$ is a continuous function over some interval of the real variable v. With the condition

$$x(v_0) = x_0 \tag{2}$$

x_0 being a given vector, (1) has a unique solution over the interval considered [4]. Integrating both sides of (1) from v_0 to v yields

$$Ax(v) - b = e^{A[g(v)-g(v_0)]}(Ax_0 - b) \tag{3}$$

where

$$g(v) \equiv \int f(v)dv \tag{4}$$

is a primitive for $f(v)$, i.e., $f(v) = dg/dv$. Thus, (1) can be written in the form

$$\frac{dx}{dv} = f(v)e^{A[g(v)-g(v_0)]}(Ax_0 - b) \tag{5}$$

Integrating again both sides from v_0 to v gives

$$x(v) = x_0 + e^{-Ag(v_0)}\sum_{k=0}^{\infty}\frac{A^k}{(k+1)!}\left[(g(v))^{k+1} - (g(v_0))^{k+1}\right](Ax_0 - b) \tag{6}$$

Two particular cases are presented.

If $f(v)$ is chosen to be $f(v) = -1$ then $g(v) = v$. Choosing $v_0 = 0$ gives $g(v_0) = 0$ and (6) becomes

$$x(v) = e^{-Av}x_0 + v\sum_{k=0}^{\infty}(-1)^k\frac{(Av)^k}{(k+1)!}b \tag{7}$$

If $f(v) = 1/v$ then $g(v) = \ln v$. With $v_O = 1$ and $g(v_o) = 0$ (6) becomes now

$$x(v) = x_o + (\ln v)\sum_{k=0}^{\infty}\frac{(A\ln v)^k}{(k+1)!}(Ax_o - b) \tag{8}$$

The solution of (1) with $f(v) = 1/v$ can also be obtained by employing a Taylor series expansion about $x_0 = 1$. Indeed,

$$x(v) = x(1) + \frac{(v-1)}{1!}x^{(1)}(1) + \frac{(v-1)^2}{2!}x^{(2)}(1) + \cdots + \frac{(v-1)^k}{k!}x^{(k)}(1) + \cdots \tag{9}$$

and evaluating the derivatives yields

$$x(v) = x_o - (1-v)\left[I + \sum_{k=1}^{\infty}\frac{(1-v)^k}{k+1}(I-A)\left(I-\frac{A}{2}\right)\cdots\left(I-\frac{A}{k}\right)\right](Ax_o - b) \tag{10}$$

where I is the identity matrix. The series in (10) is convergent for $v \in (0,2)$ but its rate of convergence is, in general, very small for v very close to zero.

Relationship with linear systems of algebraic equations

Consider now a system of equations written in matrix form as

$$Ax = b \tag{11}$$

and assume that x is a continuous function of the real variable v over a certain interval. The solution of (11) can be obtained from (3) for a v _ v_S for which

$$Ax(v_S) - b = 0 \tag{12}$$

To satisfy this condition $g(v)$ in (3) must be chosen such that

$$e^{A[g(v_s) - g(v_o)]} = 0 \tag{13}$$

For positive definite matrices A and a finite $g(v_o)$ this is achieved if $g(v)$ □ $-\infty$ for v ! v_S. Thus, the solution of (11) cannot be computed directly from (6). On the other hand, in the particular case of $f(v) = 1{=}v$, $g(v) = \ln v$ the solution of (11) can be obtained from (10) with $v = v_S = 0$,

$$x_s = x_o - \left[I + \sum_{k=1}^{\infty}\frac{1}{k+1}(I-A)\left(I-\frac{A}{2}\right)\cdots\left(I-\frac{A}{k}\right)\right](Ax_o - b) \tag{14}$$

The expression in the brackets is just the inverse of A,

$$A^{-1} = I + \sum_{k=1}^{\infty} \frac{1}{k+1}(I - A)\left(I - \frac{A}{2}\right)\cdots\left(I - \frac{A}{k}\right) \tag{15}$$

The rate of convergence of the series in (14) and (15) is very small and they cannot be practically used in numerical computation for arbitrary matrices.

$$x_s = x_o + e^{Ag(v_s)}\left(e^{Ag(v_o)} - e^{Ag(v)}\right)^{-1}\left(x(v) - x_o\right) \tag{16}$$

We note that the solution of (11) can be formally expressed from (6) as

which is valid for any $v \neq v_0$ in the interval considered. To compute x_s with (16) would require the computation of $\left(e^{Ag(v_o)} - e^{Ag(v)}\right)^{-1}$. Equation (16) will be used in Subsection 4.1.

Note. The matrix product in the terms of the series in (10), (14) and (15) can be expressed as a matrix polynomial using the relationship with the Stirling numbers of the first kind $S_{k+1}^{(m)}$ [5]

$$(I - A)\left(I - \frac{A}{2}\right)\cdots\left(I - \frac{A}{k}\right) = (-1)^k \frac{1}{k!}\sum_{m=1}^{k+1} S_{k+1}^{(m)} A^{m-1} \tag{17}$$

While each new term in such series as those in (10), (14) and (15) is calculated through a multiplication with a matrix that becomes more and more well-conditioned as k increases, the computation with the expression in (17) would require successive multiplications with the same original matrix and, for each k, a new polynomial is to be constructed and new Stirling numbers have to be generated. The formulae derived in the next two sections contain the same type of series and, therefore, are simpler and more efficient to be used for numerical computations.

3. NEW FORMULAE FOR COMPUTING MATRIX EXPONENTIALS

In this section, we derive matrix exponential expressions which contain highly convergent infinite series that allow accurate and stable numerical computations in numerous applications. They shall also be used in the next section.

3.1 Series expressions for matrix exponentials

Consider the matrix function v^A $e^{A \ln v}$ where v is a positive real variable and A a general square matrix with real number entries. By integration,

$$\int_1^v \nu^{A-I} d\nu = \left.\frac{\nu^{A}}{A}\right|_{\nu=1}^{v} = \frac{1}{A}\left(e^{A\ln v} - I\right) \tag{18}$$

For $v \in (0, 2)$ the integrand can be expanded in a power series as

$$\begin{aligned} v^{A-I} = \left[1-(1-v)\right]^{A-I} = I - \frac{(1-v)}{1!}(A-I) + \frac{(1-v)^2}{2!}(A-I)(A-2I) \\ - \frac{(1-v)^3}{3!}(A-I)(A-2I)(A-3I) + \cdots \end{aligned} \tag{19}$$

that can be integrated term by term. From (18) and (19)

$$e^{A\ln v} = I - (1-v)A\left[I + \sum_{k=1}^{\infty}\frac{(1-v)^k}{k+1}(I-A)\left(I-\frac{A}{2}\right)\cdots\left(I-\frac{A}{k}\right)\right] \tag{20}$$

which is valid for any A, positive definite or not, of arbitrary condition. One can see that, for a positive definite A and for $v \to 0$, the expression in the brackets of (20) gives a series expansion for the inverse A^{-1} (see (14) and (15)). Various expressions for the matrix exponential are obtained by giving particular values to v in (20). For example, for any $v = e^{-q}$, $q > \ln 2$, we have the series becoming less convergent as q increases above $q = 0$. On the other hand, with $v = 1/e$ in (20) and then A replaced with qA we obtain for any real number q this series being more convergent than that in (21) for greater values of q.

$$e^{-qA} = I - (1-e^{-q})A\left[I + \sum_{k=1}^{\infty}\frac{(1-e^{-q})^k}{k+1}(I-A)\left(I-\frac{A}{2}\right)\cdots\left(I-\frac{A}{k}\right)\right] \tag{21}$$

$$e^{-qA} = I - (1-e^{-1})qA\left[I + \sum_{k=1}^{\infty}\frac{(1-e^{-1})^k}{k+1}(I-qA)\left(I-\frac{qA}{2}\right)\cdots\left(I-\frac{qA}{k}\right)\right] \tag{22}$$

For any $v \in (1, 2)$, the terms in the series expressions derived from (20) have coefficients that are alternately positive and negative. With $v = e^{1/2} = 1.64872127$, e.g., and then replacing A with $2A$ we have

$$e^{A} = I + 2(e^{1/2}-1)A\left[I + \sum_{k=1}^{\infty}\frac{(-1)^k(e^{1/2}-1)^k}{k+1}(I-2A)\left(I-\frac{2A}{2}\right)\cdots\left(I-\frac{2A}{k}\right)\right] \tag{23}$$

As well, by replacing A by $-qA$ in (23) one obtains instead of (22) an expression with alternating in sign series coefficients.

Rapidly convergent series formulae

From the basic Eq. (20) we derive now formulae which contain series that have a higher rate of convergence than those presented in the previous subsection.

Firstly, it is obvious that for values of v close to 1 the series in (20) has a high rate of convergence. For instance, with $v = 1 + 10^{-q}$, $10^{-q} \ll 1$, and replacing $A \ln(1+10^{-q})$ by A we obtain

$$e^{A} = I + 10^{-q} c_q A \left[I + \sum_{k=1}^{\infty} \frac{(-1)^k 10^{-qk}}{k+1} (I - c_q A)\left(I - \frac{c_q A}{2}\right)\cdots\left(I - \frac{c_q A}{k}\right) \right] \tag{24}$$

where $c_q \equiv 1/\ln(1+10^{-q})$. This series is rapidly convergent.

Secondly, the convergence of the series in the expressions derived from (20) can further be improved by successive integrations. Indeed, integrating both sides of (20) from 1 to v, we have

$$\begin{aligned} v e^{A\ln v} = I + (1-v)(I+A)\Bigg[-I + \frac{1-v}{2}A \\ + A\sum_{k=1}^{\infty} \frac{(1-v)^{k+1}}{(k+1)(k+2)} (I-A)\left(I - \frac{A}{2}\right)\cdots\left(I - \frac{A}{k}\right) \Bigg] \end{aligned} \tag{25}$$

Integrating repeatedly we obtain the identity

$$\begin{aligned} v^p e^{A\ln v} = I - p!\sum_{k=1}^{p} \frac{(-1)^k (1-v)^k}{k!(p-k)!}\left(I + \frac{A}{p}\right)\left(I + \frac{A}{p-1}\right)\cdots\left(I + \frac{A}{p-k+1}\right) \\ + (-1)^{p+1}(1-v)^{p+1} A (I+A)\left(I + \frac{A}{2}\right)\cdots\left(I + \frac{A}{p}\right)\Bigg[\frac{1}{p+1} I \\ + p!\sum_{k=1}^{\infty} \frac{(1-v)^k}{(k+1)(k+2)\cdots(k+p+1)} (I-A)\left(I - \frac{A}{2}\right)\cdots\left(I - \frac{A}{k}\right)\Bigg], \\ p = 1, 2, \ldots; \quad v \in (0, 2) \end{aligned} \tag{26}$$

Same result is obtained by replacing A with $pI + A$ in (20). This identity contains an infinite series whose coefficients decrease rapidly as p increases. Obviously, for a given A and v, (26) generates more efficient computational formulae than those in the previous subsection. As before, for $v \in (1, 2)$ the infinite series have coefficients that alternate in sign. For example, with $v = e^{1/2}$ in (26) and then replacing A by $2A$ and p by $2p$ we have instead of (23)

$$
\begin{aligned}
e^{A} = e^{-p}\Bigg\{ & I + (2p)!\sum_{k=1}^{2p} \frac{(e^{1/2}-1)^k}{k!(2p-k)!}\left(I+\frac{2A}{2p}\right)\left(I+\frac{2A}{2p-1}\right)\cdots\left(I+\frac{2A}{2p-k+1}\right) \\
& + 2(e^{1/2}-1)^{2p+1} A(I+2A)\left(I+\frac{2A}{2}\right)\cdots\left(I+\frac{2A}{2p}\right)\Bigg[\frac{1}{2p+1}I \\
& + (2p)!\sum_{k=1}^{\infty} \frac{(-1)^k(e^{1/2}-1)^k}{(k+1)(k+2)\cdots(k+2p+1)}(I-2A)\left(I-\frac{2A}{2}\right)\cdots\left(I-\frac{2A}{k}\right)\Bigg]\Bigg\}, \\
& p = 1,2,\ldots
\end{aligned}
\tag{27}
$$

Taking $v = 1 + 10^{-q}$, with $q>0$ and $c_q \equiv 1/\ln(1 + 10^{-q})$, (26) gives (compare with (24))

$$
\begin{aligned}
e^{A} = (1+10^{-q})^{-p}\Bigg\{ & I + p!\sum_{k=1}^{p} \frac{10^{-qk}}{k!(p-k)!}\left(I+\frac{c_qA}{p}\right)\left(I+\frac{c_qA}{p-1}\right)\cdots\left(I+\frac{c_qA}{p-k+1}\right) \\
& + 10^{-(p+1)q} c_q A(I+c_qA)\left(I+\frac{c_qA}{2}\right)\cdots\left(I+\frac{c_qA}{p}\right)\Bigg[\frac{1}{p+1}I \\
& + p!\sum_{k=1}^{\infty} \frac{(-1)^k 10^{-qk}}{(k+1)(k+2)\cdots(k+p+1)}(I-c_qA)\left(I-\frac{c_qA}{2}\right)\cdots\left(I-\frac{c_qA}{k}\right)\Bigg]\Bigg\}, \\
& p = 1,2,\ldots
\end{aligned}
\tag{28}
$$

Notice that, in the new formulae derived from (26) the infinite series are very rapidly convergent, with their rate of convergence increasing when the parameter p increases. Highly accurate numerical results can be generated with only a small number of terms retained in the infinite series of these formulae (see Section 4).

Note. All the formulae presented in this section remain valid if A is changed in A. Obviously, in all these expressions A can be replaced by a real number and the identity matrix I by 1, yielding a few novel identities and summation formulae for series of real numbers. Also, the expression in the brackets of (26) for v →0 is just $(I + A/p)^{-1}/p$ if $I+A/p$ is positive definite and, thus, we obtain another identity, i.e.,

$$
\begin{aligned}
& p!\sum_{k=1}^{p} \frac{(-1)^k}{k!(p-k)!}\left(I+\frac{A}{p}\right)\left(I+\frac{A}{p-1}\right)\cdots\left(I+\frac{A}{p-k+1}\right) \\
& \quad = I + \frac{(-1)^{p+1}}{p} A(I+A)\left(I+\frac{A}{2}\right)\cdots\left(I+\frac{A}{p-1}\right), \quad p = 2,3,\ldots
\end{aligned}
\tag{29}
$$

which reduces for $A = 0$ to an elementary binomial sum.

4. SOLUTION OF GENERAL LINEAR SYSTEMS

In what follows, we apply the matrix exponential formulae from the previous section and present a new iteration procedure and a matrix product formula for the solution of large systems of linear algebraic equations.

An iterative method

Equation (16) can be written in the form

$$x(v) = x_o + \left(I - e^{A[g(v) - g(v_o)]}\right)(x_S - x_o) \tag{30}$$

where $x_S = x(v_S)$ is the solution of (11), $x_o = x(v_o)$, A is a positive definite matrix and $g(v)$ is a function of the real variable v such that $g(v) \to -\infty$ for $v \to v_s$ (see Subsection 2.2). To get $x(v)$ for values of v very close to v_S we choose an adequate $g(v)$ and a formula for the matrix exponential from Section 3. When $g(v_o) = 0$, applying (22) for instance gives

$$\begin{aligned} x(v) = x_o + (1 - e^{-1})g(v)\Bigg[I + \sum_{k=1}^{\infty} \frac{(1-e^{-1})^k}{k+1} (I + Ag(v)) \\ \times \left(I + \frac{A}{2} g(v)\right) \cdots \left(I + \frac{A}{k} g(v)\right) \Bigg] (Ax_o - b) \end{aligned} \tag{31}$$

If $g(v) \equiv \ln v$, with $v_0 = 1$ and $v_s = 0$, we compute $x(e^{-N})$ for $N \gg 1$ which is closer to the solution x_S,

$$\begin{aligned} x(e^{-N}) = x_S^{(1)} = x_S^{(0)} - N(1 - e^{-1})\Bigg[I + \sum_{k=1}^{\infty} \frac{(1-e^{-1})^k}{k+1} (I - NA) \\ \times \left(I - \frac{N}{2} A\right) \cdots \left(I - \frac{N}{k} A\right) \Bigg] (Ax_S^{(0)} - b) \end{aligned} \tag{32}$$

where $x_S^{(0)} \equiv x_o$. This equation is applied iteratively by replacing $x_S^{(1)}$ and $x_S^{(0)}$, respectively, with $x_s^{(i)}$ and $x_s^{(i-1)}$, satisfies (11) with a desired accuracy.

To evaluate the amount of computation necessary to obtain the solution of (11) with a certain accuracy, let us take N such that $N||A||= 10$ when one needs about 30 terms in the infinite series, i.e., 30 matrix-vector multiplications. The number of iterations increases with the condition number of A. To see this and to determine the corresponding number of iterations, consider (11) with b=0 and A replaced with a diagonal matrix whose entries are positive numbers, the greatest of these being 1, and whose condition is the same as that of A. The solution of this system is $x_S = 0$ and the components of the solution of (1), with $f(v)=1/v$, are $x_o k v^{\lambda k}$ where λ_k, k=1,2,...,n, are the entries in the diagonal matrix. In order to make the magnitudes of all these components at least 1%, e.g., of the corresponding magnitudes of the initial components, one needs no iteration if the condition number is less than 2, but 5 and, respectively, 46 iterations are needed if the condition number is 10 and 100.

What is remarkable in the iterative method based on (32) is that, for matrices with same condition number and same norm, the number of iterations required is the same, independently of the size of the matrices. Considering approximately $2n^2$ arithmetic operations for one matrix-vector multiplication, where n is the number of equations and unknowns in (11), the total number of arithmetic operations required is, thus, proportional to only n^2. In the examples given above one has to perform, respectively, $60n^2$, $300n^2$ and $2760n^2$ arithmetic operations. Assuming only $2n^3/3$ arithmetic operations for the Gaussian elimination procedure, the method presented in this subsection requires less computation for the same examples if, respectively, n>90, n>450 and n>4140. One can also notice that the application of Eq. (32) leads to the actual solution of (11) independently of the small error introduced in the computation at each iteration.

A matrix product formula

The original general system (11) is replaced with an equivalent system such that its solution is obtained in terms of matrix exponentials for which highly convergent and accurate series formulae have been derived in Section 3.

Namely, instead of (11) we use the system

$$\left(I-e^{-\alpha A}\right)x=b_{\alpha} \tag{33}$$

where α is a real scalar to be chosen, α ≠ 0, and

$$b_{\alpha}=\alpha\left[1-\frac{(\alpha A)}{2!}+\frac{(\alpha A)^2}{3!}-\frac{(\alpha A)^3}{4!}+\cdots\right]b \tag{34}$$

Assuming A to be positive definite, α is taken positive. Then, since $\|e^{-\alpha A}\|<1$ for a normal matrix, the solution can be expressed as

$$x_s=\left(I-e^{-\alpha A}\right)^{-1}b_{\alpha}=\left[\sum_{k=0}^{\infty}e^{-k\alpha A}\right]b_{\alpha} \tag{35}$$

with the norm of the matrix exponentials decreasing when k increases [6]

$$\left\|e^{-k\alpha A}\right\|<e^{-k\alpha\lambda} \tag{36}$$

where λ is the smallest eigenvalue of A. b_α can be accurately computed by using instead of (34) an equivalent expression, for instance (see (22))

$$b_\alpha = (1-e^{-1})\alpha \left[I + \sum_{k}^{\infty} \frac{(1-e^{-1})^k}{k+1} (I-\alpha A)\left(I-\frac{\alpha A}{2}\right)\cdots\left(I-\frac{\alpha A}{k}\right)\right] b \tag{37}$$

If the infinite series in (35) is truncated to $k = N_S$ the rest of the series has a norm

$$\left\| R_{N_s} \right\| < \frac{e^{-(N_s+1)\alpha\lambda}}{1-e^{-\alpha\lambda}} \tag{38}$$

Much less numerical computation (see below) is needed if the infinite series in (35) is transformed into an infinite product using the identity [5]

$$(1-u)^{-1} = \prod_{k=0}^{\infty} (1+u^{2^k}), \quad 0<u<1 \tag{39}$$

which is also valid for matrices whose norm is less than 1. Thus, (35) becomes

$$x_s = \left[\prod_{k=0}^{\infty} (1+e^{-2^k \alpha A}) \right] b_\alpha \tag{40}$$

with the norm of the exponentials $e^{-2k\alpha A}$ decreasing very rapidly when k increases. Truncating the infinite product to $k = N_P$, i.e., N_P+1 factors, leaves a remaining factor

$$F_{N_p} = \prod_{k=N_p+1}^{\infty} (I+e^{-2^k \alpha A}) \tag{41}$$

whose norm is

$$\left\| F_{N_p} \right\| < \prod_{k=N_p+1}^{\infty} (1-e^{-2^k \alpha\lambda}) \quad \frac{1}{1-e^{-\alpha\lambda}} \frac{1}{\prod_{k=0}^{N_p} (1+e^{-2^k \alpha\lambda})} \tag{42}$$

Let us compare the maximum value of the norm of the truncated matrix in the brackets of (35) and (40) with that of the corresponding untruncated matrix in order to get a rough estimate of the numbers N_S and N_P of matrix exponentials involved in the numerical computation to achieve a certain accuracy. This will also allow to estimate the total number of matrix-vector multiplications necessary to obtain the solution. The ratio of the maximum norm of the truncated matrix to the maximum norm of the untruncated matrix in the brackets of (35) and (40) is, respectively, (see (38) and (42))

$$\rho = 1 - e^{-(N_S+1)\alpha\lambda} \tag{43}$$

and

$$\rho = (1 - e^{-\alpha\lambda}) \prod_{k=0}^{N_P} (1 + e^{-2^k \alpha\lambda}) \tag{44}$$

To illustrate the computation complexity when using (35) or (40), assume that $||A||=1$ and $\alpha = 20$. If ρ is imposed to be $\approx 0{:}99$, for example, one needs $N_S = 2$, i.e., three terms in (35) and $N_P = 1$, i.e., two factors in (40) if $\lambda = 10^{-1}$. If $\lambda = 10^{-2}$ these numbers increase to $N_S = 23$ and $N_P = 4$, and when $\lambda = 10^{-3}$ one gets $N_S = 230$, but N_P only increases to $N_P = 7$. It is clear that applying the formula (40) the number of exponentials needed in the numerical computation is much smaller than that if formula (35) would be applied. For all the matrix exponentials involved in the numerical computation we use the formula (28) containing a highly convergent series, such that

$$\begin{aligned} e^{-\alpha A} = (1-10^{-q})^{-p} \Bigg\{ & I + p! \sum_{k=1}^{p} \frac{10^{-qk}}{k!(p-k)!} \left(I - \frac{c_q \alpha A}{p}\right)\left(I - \frac{c_q \alpha A}{p-1}\right) \cdots \left(I - \frac{c_q \alpha A}{p-k-1}\right) \\ & - 10^{-(p+1)q} c_q \alpha A (I - c_q \alpha A)\left(I - \frac{c_q \alpha A}{2}\right) \cdots \left(I - \frac{c_q \alpha A}{p}\right) \Bigg[\frac{I}{p+1} \\ & + p! \sum_{k=1}^{\infty} \frac{(-1)^k 10^{-qk} k!}{(k+p+1)!} (I + c_q \alpha A)\left(I + \frac{c_q \alpha A}{2}\right) \cdots \left(I + \frac{c_q \alpha A}{k}\right) \Bigg] \Bigg\}, \\ & p = 1, 2, \ldots \end{aligned} \tag{45}$$

where $q > o$ and $c_q \equiv 1/\ln(1\ 10^{-q})$. With $\alpha||A||=20$ and choosing $q = 1$ and $p = 10$, for instance, $e^{-\alpha A}$ is determined accurately by retaining 50 terms in the infinite series and, thus, to multiply $e^{-\alpha A}$ with a vector one needs 71 matrix-vector multiplications. To compute x_s from (40) one has to use repeatedly the multiplication of e^{-20A} with a vector. For a matrix A with $\lambda = 10^{-1}$ one has to retain two factors in (40) and, thus, to multiply e^{-20A} and $e^{-2\times 20A}$ with a vector. This means to use repeatedly 3 times the multiplication of e^{-20A} with a vector which requires, therefore, $3\times 71 = 213$ matrix-vector multiplications. When $\lambda = 10^{-2}$ the infinite product in (40) is truncated at $k = N_P = 4$ and this requires the multiplication of $e^{-2k\times 20A}$, $k = 0, 1, 2, 3, 4$, with a vector, i.e., to use repeatedly 31 times the multiplication of e^{-20A} with a vector, for a total of $31\times 71=2201$ matrix-vector multiplications. We also have to add the matrix-vector multiplications required to compute b_α in (37). A very accurate result for b_α when $\alpha = 20$ can be achieved by applying four times the series in the brackets of (37) for $\alpha = 5$, each time retaining 30 terms. This requires a total of about 120 multiplications of a matrix $I - 5A/k$ with a vector. In all the matrix-vector multiplications involved when applying (45), the matrices are in the form $I \pm c_q \alpha A/k$ and become better and better conditioned as k increases.

Adding up the number of arithmetic operations involved shows that, with respect to the classical Gaussian elimination method, the procedure presented in this subsection is advantageous for very large systems (11). Namely, assuming same accuracy and only $2n^3/3$ arithmetic operations for the Gaussian elimination, with the data given above, one has to have $n>3\times(213+120)=999$ equations and unknowns if $\lambda = 10^{-1}$ and $n>3\times(2201+120) = 6963$ equations and unknowns if $\lambda = 10^{-2}$ for the proposed method to be more advantageous. For a given $\alpha||A||$, one application of $e^{-\alpha A}$ requires a determined finite number of matrix-vector multiplications, independently of the size of A. It is remarkable, as for the iterative method in the previous subsection, that for a given condition of A, one has to apply $e^{-\alpha A}$ a well-determined number of times and, thus, the total number of arithmetic operations necessary to compute the solution with an imposed accuracy is proportional to only n^2.

It should be noted that, since the infinite series in the expression (45) is truncated and thus determined with a finite accuracy, the accuracy of the solution x_s becomes compromised after a too big a number of matrix exponential-vector multiplications. This is why, the worse conditioned systems (11) should be appropriately preconditioned. Practically, the computation with (40) is continued factor by factor and the accuracy of x is checked after each step.

5. SOLUTION OF GENERAL LINEAR SYSTEMS BY NUMERICAL INTEGRATION OF DIFFERENTIAL EQUATIONS

In this section, we introduce first order differential equations whose numerical integration allows to efficiently find the solution of linear systems of algebraic equations. Differential equations of the type of those in (1), with $f(v)=-1$ or $f(v) = 1/v$, cannot be used for this purpose due to the fact that the first and higher order derivatives of $x(v)$ tend to infinite values as x tends to the solution x_s of (11) (see Section 2).

Here below, we construct ordinary differential equations for $x(v)$ which satisfy the condition that the first few derivatives are finite when $x(v)$ tends to x_s and, therefore, are particularly useful for an accurate computation of x_s. Let us consider the system (11) with a symmetric positive definite matrix. A quadratic functional

$$F(x)=\frac{1}{2}x^{T}Ax-x^{T}b+\frac{1}{2}b^{T}A^{-1}b \tag{46}$$

is associated with (11) [6] whose minimum value is $F(x_s) = 0$. Define now a real variable v, $v \geq 0$, such that

$$F(x)=v^{r} \tag{47}$$

where r is a real number to be chosen, $r>0$, with $v = 0$ corresponding to the solution $x = x_s$ and $v = v_0$ to an initial point x_o, $F(x_o) = v_0^r$. Then,

$$\frac{dF}{dv}=rv^{r-1}=(Ax-b)^{T}\frac{dx}{dv} \tag{48}$$

and, thus,

$$\frac{dx}{dv}=rv^{r-1}\frac{Ax-b}{\left\|Ax-b\right\|^{2}} \tag{49}$$

This is the differential equation to be integrated from $v = v_o$ to $v = v_s = 0$. The second derivative of x is obtained in the form

$$\begin{aligned}\frac{d^{2}x}{dv^{2}}=r(r-1)v^{r-2}\frac{Ax-b}{\left\|Ax-b\right\|^{2}}+(rv^{r-1})^{2}\frac{1}{\left\|Ax-b\right\|^{4}}\Bigg\{A\\-2\frac{\left[(Ax-b)^{T}A(Ax-b)\right]}{\left\|Ax-b\right\|^{2}}\Bigg\}(Ax-b)\end{aligned} \tag{50}$$

Higher order derivatives can be worked out if needed.

In order to see the behaviour of the derivatives close to the solution x_s, Eqs. (49) and (50) are rewritten as

$$\frac{dx}{dv}=r\frac{(F(x))^{1-\frac{1}{r}}(Ax-b)}{\left|Ax-b\right|^{2}} \tag{51}$$

and

$$\frac{d^2x}{dv^2} = r(r-1)\frac{(F(x))^{1-2/r}\cdot(Ax-b)}{\|Ax-b\|^2} + r^2\frac{(F(x))^{2-2/r}}{\|Ax-b\|^4}\left\{A - 2\frac{\left[(Ax-b)^T A(Ax-b)\right]}{\|Ax-b\|^2}\right\}(Ax-b) \tag{52}$$

with $F(x)$ in (46) put in the form

$$F(x) = \frac{1}{2}(Ax-b)^T(x-x_S) \tag{53}$$

Notice that as x tends to x_s, when $\|Ax-b\| \equiv \varepsilon$ tends to zero,

$$\left\|\frac{dx}{dv}\right\| = K_1(v)\varepsilon^{1-2/r}, \quad \left\|\frac{d^2x}{dv^2}\right\| = K_2(v)\varepsilon^{1-4/r} \tag{54}$$

where $K_1(v)$ and $K_2(v)$ are finite when $v \to 0$. Therefore, as

$x \to x_S$, $\|dx/dv\| \to 0$ if $r>2$ and $\|d^2x/dv^2\| \to 0$ if $r>4$.

Another differential equation we present here is

$$\frac{dx}{ds} = \frac{Ax-b}{\|Ax-b\|} \tag{55}$$

with the second derivative

$$\frac{d^2x}{ds^2} = \frac{1}{\|Ax-b\|^2}\left\{A - \frac{\left[(Ax-b)^T A(Ax-b)\right]}{\|Ax-b\|^2}\right\}(Ax-b) \tag{56}$$

In this case, always

$$\left\|\frac{dx}{ds}\right\| = 1 \tag{57}$$

even for $x \to x_s$, but the second derivative tends to an infinite value when $x \to x_s$

$$\left\|\frac{d^2x}{ds^2}\right\| \sim K(s)\varepsilon^{-1} \tag{58}$$

where $K(s)$ is finite and $\varepsilon \equiv \|Ax - b\|$. The relationship between the differentials of the variables v and s in (49) and (55) is

$$ds = \frac{r(F(x))^{1-\frac{1}{r}}}{\|Ax - b\|}dv \tag{59}$$

and for $x \to x_s$ we have (see (54))

$$\frac{ds}{dv} \sim K_1(v)\varepsilon^{1-\frac{2}{r}} \tag{60}$$

The differential Eqs. (49) in v and (55) in s require practically the same amount of computation for their right-hand sides, i.e., one matrix-vector multiplication. The first derivatives dx/dv for $r = 2$ and dx/ds remain finite when x tends to x_s, while the second derivatives increase to infinite values as in (54) and (58). For $r = 4$ the second derivative in (52) remains finite when x tends to x_s (see (54)), while the first derivative and the ratio ds/dv tend to zero as in (54) and (60), respectively. If $r>4$ even $\|d^2x/dv^2\|$ tends to zero as in (54).

Equations (49) and (55) can be integrated by classical numerical methods. Since we are not looking for an accurate solution of these equations all along from x_o to x_s but for finding accurately the final value $x = x_s$, we can use a lower order method, for instance, even Euler's method [7]. This yields an approximate value of x_s which is to be used as initial point for repeating the numerical integration procedure. As we get closer to the solution x_s, we decrease the step size in order to reduce the error. In the case of Euler's method the error is determined in terms of the norm of the second derivative. Higher order numerical integration methods can also be used in order to increase the computation efficiency.

To find a starting point for the integration procedure which is reasonably close to the solution point, one can minimize $F(x)$ in (46) along the normal direction, followed by a minimization of the distance to the solution point x_s along the direction of the normal to F [8]. These two preliminary steps are repeated a few times as needed.

Numerical experiments have been performed applying Euler's method to (49) for $r = 2$, $r = 4$ and $r = 8$, and to (55). Systems (11) of various sizes have been automatically generated and the differential Eqs. (49) for $r = 2$ and $r = 4$, and (55) have produced results with the least amount of computation when imposing an accuracy of 1%.

For matrices which are not symmetric positive definite, (46) is replaced with $F(x) = 1/2\ (Ax - b)^T\ (Ax - b)$:

6. CONCLUSIONS

A special type of matrix series are used in Section 2 to express the relationship between some first order ordinary differential equations and systems of linear algebraic equations and, also, in Section 3 to derive efficient formulae for matrix exponentials that allow accurate and stable numerical computations in various applications. The main feature of these series consists in the fact that, starting with their first term which is already a matrix substantially better conditioned than the original problem matrix, each of the subsequent terms is obtained through a multi- plication with a better and better conditioned matrix that tends to the identity matrix. The new matrix exponential formulae contain very rapidly convergent series and can be applied to general, arbitrarily conditioned, positive definite or not matrices. They are used in Section 4 for two new methods of solution for general linear algebraic systems. One is an iterative method which corresponds to the solution of the differential Eq. (1) with $f(v) = 1/v$: It is based on the exact analytical expressions (30)–(32) that always yield results converging finally to the exact solution of the system (11). In a second method, the original algebraic system (11) is replaced with an equivalent system containing a matrix exponential $e^{-\alpha A}$ such that instead of inverting the system matrix A we have now to invert $I- e^{-\alpha A}$. The exact analytical solution is obtained in the form of a series of matrix exponentials which is transformed into an infinite matrix product in order to reduce substantially the necessary amount of computation. It should be remarked that, since the number of matrix-vector multiplications required for the application of one matrix exponential-vector multiplication only depends on the norm of the matrix while the number of matrix exponential-vector multiplications depends on the condition of the system matrix, the total number of arithmetic operations needed to achieve an imposed accuracy when applying each of the two methods is practically proportional to n^2, where n is the dimension of the matrix. The two methods require a comparable total amount of computation. It is also remarkable that for both methods the necessary amount of computation can be roughly predicted before- hand in terms of the system size, the system condition and the desired accuracy.

AT A GLANCE

The solution of specific differential equations is represented using a distinct form of matrix series and is directly connected to the solution of overall systems of algebraic equations. The derivation of efficient equations for matrix exponentials involves the use of rapidly convergent series of the same type. These two new solution strategies, namely an iterative method and a method based on an accurate matrix product formula, are crucial for solving large linear systems. An analysis is conducted on the computational complexity of these two methods. It is established that for both methods, the number of matrix exponential-vector multiplications needed to achieve a desired level of accuracy may be predetermined based on the system state. The total number of arithmetic operations required is almost directly proportional to the square of the matrix dimension, denoted as n. All the series in the reported findings share a common characteristic: they begin with a first term that is already well-conditioned, and each subsequent term is calculated by multiplying it with an even better conditioned matrix. This process leads to a rapid convergence towards the identity matrix. This significantly enhances the stability of the numerical computation. This paper presents a highly effective technique that utilizes numerical integration to solve a specific type of differential equations. This method can be used to systems that are even ill-conditioned.

REFERENCES

1. Ciric IR. Rapidly convergent matrix series for solving linear systems of equations. In: Proceedings of the 17th International Symposium on Antenna Technology and Applied Electromagnetics (ANTEM); 10-16 July 2016; Montréal, Canada: IEEE;W 2016. DOI: 10.1109/ANTEM.2016.755022/ 978-1-4673-8478-0
2. Lanczos C. Applied Analysis. New York: Dover Publications, Inc.; 1988. ISBN: 0-486-65656-X (Paperback). CIP: 88-3961
3. Moler CB, Van Loan CF. Nineteen dubious ways to compute the exponential of a matrix. SIAM Review. 1978;**20**:801-836. DOI: 10.1137/ S00361445024180
4. Hurewicz W. Lectures on Ordinary Differential Equations. New York: Dover; 1990. ISBN: 0-486-66420-I (Paperback). QA372.H93
5. Gradshteyn IS, Ryzhik IM. Table of Integrals, Series and Products. 5th ed. New York: Academic Press; 1994. ISBN: 0-12-294755-X. QA55.G6613
6. Golub GH, Van Loan CF. Matrix Computations. 2nd ed. Baltimore: Johns Hopkins; 1989. ISBN: 0-8018-3772-3, 0-8018-3739-1 (Paperback). 88-4504
7. Burden RL, Faires JD. Numerical Analysis. 5th ed. Boston: PWS-KENT; 1993. ISBN: 0-534-93219-3. CIP: 92-32192

8. Ciric IR. A geometric property of the functional associated with general systems of algebraic equations. In: Proceedings of the International Symposium on Antenna Technology and Applied Electromagnetics and Canadian Radio Sciences Conference (ANTEM/URSI); 17-19 July 2006; Montréal, Canada: IEEE; 2006. p. 311-315. ISBN: 978-0-9738425-1-7

03 FIXED POINT THEOREMS OF A NEW GENERALIZED NONEXPANSIVE MAPPING

1. INTRODUCTION: AN OVERVIEW

In this paper, E is a Banach space, C is a nonempty closed convex subset of E, and $Fix(T)=x\in C:Tx=x$.

Definition 1. T is contraction mapping if there is $r\in[0,1)$

$$\|Tx-Ty\| \le r\|x-y\| \text{ for all } x\ \ y\in C.$$

Definition 2. T is nonexpansive mapping if

$$\|Tx-Ty\| \le \|x-y\| \text{ for all } x\ \ y\in C.$$

Definition 3. T is quasinonexpansive mapping if

$$\|Tx-Ty\| \le \|x-y\| \text{ for all } x\in C, y\in F(T).$$

Definition 4. $T:C\to C$ is a T-(D_a) mapping on a subset C, if there is

$a\in\left(\frac{1}{2},1\right)$, $\|Tx-Ty\|\le\|x-y\|$ for all $\alpha\in[a,1], x\in C, y\in C(T,x,\alpha)$, where $C(T,x,\alpha)=\{y\in C|y=(1-\alpha)p+\alpha Tp, p\in C, \|Tp-p\|\le\|Tx-x\|\}$.

In 2008 Suzuki [1] defined a mapping T in Banach space: $\frac{1}{2}\|Tx-Ty\|\le\|x-y\|$ implies $\|Tx-Ty\|\le\|x-y\|$. And T is said to satisfy condition (C). Suzuki [1] showed that the mapping satisfying condition (C) is weaker than nonexpansive mapping and stronger than quasinonexpansive mapping.

Suzuki [1] proved the theorem T is a mapping in Banach space, T satisfies condition (C), and $\{x_n\}$ is the sequence defined by the iteration process:

$$\begin{cases} x_1 = x\in C, \\ x_{n+1} = (1-\alpha_n)x_n + \alpha_n Tx_n, \end{cases} \tag{1}$$

then $\{x_n\}$ converges to a fixed point of T.

Suzuki [1] gave this convergence theorem in an ordinary Banach space, and the mapping satisfying condition (*C*) is weaker than nonexpansive mapping.

In 2016, Thakur [2] proved the theorem *T* is a mapping in uniformly convex Banach space, *T* satisfies condition (*C*), and $\{x_n\}$ is the sequence defined by iteration process:

$$\begin{cases} x_1 = x \in C, \\ x_{n+1} = Ty_n, \\ y_n = Tz_n, \\ z_n = (1-\alpha_n)x_n + \alpha_n Tx_n, \end{cases} \tag{2}$$

then $\{x_n\}$ converges to a fixed point of *T*.

Thakur [2] claimed that the rate of iteration is fastest of known iterations. However, the disadvantage is that their results must be in uniformly convex Banach space, instead of the ordinary Banach space.

The aim of this article is there exists a generalized nonexpansive mapping, which makes the sequence generated by Thakur's iteration converge to a fixed point in a general Banach space.

The following propositions are obvious:

Proposition 1. If T is nonexpansive, then T satisfies condition (D_a).

Proposition 2. If T is T-(D_a) mapping, then T is quasinonexpansive.

Proposition 3. Suppose T: C→C is a T-(D_a) mapping. Then, for x,y ∈C:

(1) $\|T^2x-Tx\| \le \|Tx-x\|$ for all x∈C.

(2) $\|T^2x-Ty\| \le \|Tx-y\|$ or $\|T^2y-Tx\| \le \|Ty-x\|$ for all x,y ∈C.

Proof:

(1) Since $\|Tx-x\| \le \|Tx-x\|$, $Tx \in C(T,x,1)$, we have $\|T^2x-Tx\| \le \|Tx-x\|$

(2) For all x,y ∈C, $\|Tx-x\| \le \|Ty-y\|$ or $\|Ty-y\| \le \|Tx-x\|$.

Then $Tx \in C(T,y,\alpha)Tx\in$ or $Ty \in C(T,x,\alpha)Ty\in$.

It follows that $\|T^2x-Ty\| \le \|Tx-y\|$ or $\|T^2y-Tx\| \le \|Ty-x\|$.

Example 1

$$Tx = \begin{cases} \begin{pmatrix} 1.1 & x_2 \\ x_3 & x_4 \end{pmatrix}, & x_1 = 3, \\ \begin{pmatrix} 0 & x_2 \\ x_3 & x_4 \end{pmatrix}, & x_1 \neq 3, \end{cases}$$

where

$$x = \begin{pmatrix} x_1 & x_2 \\ x_3 & x_4 \end{pmatrix}, x_1 \in [0,3], x_2 \in [0,0.01], x_3 \in [0,0.01], x_4 \in [0,0.01].$$

$$\|x\|_1 = \max\{|x_1| + |x_3|, |x_2| + |x_4|\}$$

Set

$$x = \begin{pmatrix} 3 & 0 \\ 0 & 0 \end{pmatrix}$$

and

$$y = \begin{pmatrix} 2.5 & 0 \\ 0 & 0 \end{pmatrix}$$

We see that

$$\|Tx - Ty\|_1 = 1.1 > \|x - y\|_1.$$

Hence, T is not a nonexpansive mapping.

To verify that T is a T-(D_a) mapping, consider the following cases:

Case 1:

$$\alpha \in \left[\frac{11}{19}, 1\right], x = \begin{pmatrix} x_1 & x_2 \\ x_3 & x_4 \end{pmatrix}, x_1 \neq 3.$$

$$y = \begin{pmatrix} y_1 & y_2 \\ y_3 & y_4 \end{pmatrix} \in C(T, x, \alpha),$$

then $y_1 \neq 3$. We have

$$\|Ty - Tx\| = \|\begin{pmatrix} 0 & y_2 - x_2 \\ y_3 - x_3 & y_4 - x_4 \end{pmatrix}\| \leq \|\begin{pmatrix} y_1 - x_1 & y_2 - x_2 \\ y_3 - x_3 & y_4 - x_4 \end{pmatrix}\| = \|y - x\|$$

Case 2:

$$\alpha \in \left[\frac{11}{19}, 1\right], x = \begin{pmatrix} x_1 & x_2 \\ x_3 & x_4 \end{pmatrix}, x_1 = 3.$$

$$y = \begin{pmatrix} y_1 & y_2 \\ y_3 & y_4 \end{pmatrix} \in C(T, x, \alpha),$$

then $y_1 \in [0, 1.9]$ We have

$$\|Ty - Tx\| = \|\begin{pmatrix} 1.1 & y_2 - x_2 \\ y_3 - x_3 & y_4 - x_4 \end{pmatrix}\| \leq 1.11 \leq \|y - x\|$$

Hence, T is a T-(D_a) mapping, and T is not nonexpansive.

2. FIXED POINT

In this section, we prove convergence theorems for fixed point of the T-(D_a) mapping in Banach space.

Lemma 1. Let *C* be bounded convex subset of a Banach space *B*. Assume that T:C→C: is T-(D_a) mapping and $\{x_n\},\{y_n\},\{z_n\}$ are sequences generated by iteration:

$$\begin{cases} x_1 = x \in C, \\ x_{n+1} = Ty_n, \\ y_n = Tz_n, \\ z_n = (1-\alpha_n)x_n + \alpha_n Tx_n, \end{cases} \tag{3}$$

where $\frac{1}{2} < a \leq \alpha_n \leq b < 1$. Then

(1) $\|Tx_{n+1}-x_{n+1}\| \leq \|Ty_n-y_n\| \leq \|Tz_n-z_n\| \leq \|Tx_n-x_n\|$

(2) $\lim_{n\to\infty}\|Tx_n-x_n\|=\lim_{n\to\infty}\|Ty_n-y_n\|=\lim_{n\to\infty}\|Tz_n-z_n\|=r\geq 0$

Proof: (1) From Proposition 3 and $z_n=(1-\alpha n)x_n+\alpha_n Tx_n$, we have

$$\begin{aligned} \|Tx_{n+1} - x_{n+1}\| &\leq \|T^2y_n - Ty_n\| \\ &\leq \|Ty_n - y_n\| = \|T^2z_n - Tz_n\| \\ &\leq \|Tz_n - z_n\| \\ &= \|Tz_n - Tx_n + (1-\alpha_n)(Tx_n - x_n)\| \\ &\leq \|z_n - x_n\| + (1-\alpha_n)\|Tx_n - x_n\| \\ &= \|Tx_n - x_n\|. \end{aligned}$$

(2) From (1), we have $0\leq\|Tx_{n+1}-x_{n+1}\|\leq\|Tx_n-x_n\|$. So $\lim_{n\to\infty}\|Tx_n-x_n\|=r\geq 0$. Now, we have $\lim_{n\to\infty}\|Tx_n-x_n\|=\lim_{n\to\infty}\|Ty_n-y_n\|=\lim_{n\to\infty}\|Tz_n-z_n\|=r\geq 0$.

Lemma 2. Assume that T:C→C is a T-(D_a) mapping and $\{x_n\},\{y_n\},\{z_n\}$ are sequences generated by iteration (3$\frac{1}{2} < a \leq \alpha_n \leq b < 1$. Let $\{u_m\}$ satisfy $u_{3n-2} = x_n, u_{3n-1}=z_n, u_{3n}=y_n$. Then, for all n≥1,p≥1

$$\begin{aligned} &\left(1+\sum_{k=3n-2}^{3n+p-3}\beta_k\right)\|Tu_{3n-2}-u_{3n-2}\| \leq \|Tu_{3n-2+p}-u_{3n-2}\| \\ &+\left(\prod_{k=n}^{n+p-1}\frac{2}{1-\alpha_k}\right)(\|Tu_{3n-2}-u_{3n-2}\|-\|Tu_{3n-2+3p}-u_{3n-2+3p}\|), \end{aligned} \tag{4}$$

where

$$\beta_k = \begin{pmatrix} \alpha_n & k = 3n-2 \\ 1 & k \neq 3n-2 \end{pmatrix}$$

Proof: From **Lemma 1**, we have

$$\begin{aligned}
&\|Tx_{n+1} - x_{n+1}\| \\
&\leq \|Tz_n - z_n\| \\
&= \|Tz_n - (1-\alpha_n)x_n - \alpha_n Tx_n\| \\
&\leq (1-\alpha_n)\|Tz_n - x_n\| + \alpha_n\|Tz_n - Tx_n\| \\
&\leq (1-\alpha_n)\|Tz_n - x_n\| + \alpha_n\|z_n - x_n\| \\
&= (1-\alpha_n)\|Tz_n - x_n\| + \alpha_n^2\|Tx_n - x_n\|.
\end{aligned}$$

So, for p=1 and all $n \geq 1$

$$\begin{aligned}
&(1+\beta_{3n-2})\|Tu_{3n-2} - u_{3n-2}\| \\
&= (1+\alpha_n)\|Tx_n - x_n\| \\
&\leq \|Tx_n - x_n\| + \left(\frac{1}{1-\alpha_n}\right)(\|Tx_n - x_n\| - \|Tx_{n+1} - x_{n+1}\|) \\
&= \|Tu_{3n-1} - u_{3n-2}\| + \left(\frac{1}{1-\alpha_n}\right)(\|Tu_{3n-2} - u_{3n-2}\| - \|Tu_{3n+1} - u_{3n+1}\|) \\
&\leq \|Tu_{3n-1} - u_{3n-2}\| + \left(\frac{2}{1-\alpha_n}\right)(\|Tu_{3n-2} - u_{3n-2}\| - \|Tu_{3n+1} - u_{3n+1}\|).
\end{aligned}$$

(4) holds.

We make the inductive assumption that (4) holds for a given p>1 and all n>0 and obtain, upon replacing n with n+1

$$\begin{aligned}
&\left(1+\sum_{k=3n+1}^{3n+p}\beta_k\right)\|Tu_{3n+1} - u_{3n+1}\| \leq \|Tu_{3n+1+p} - u_{3n+1}\| \\
&+\left(\prod_{k=n+1}^{n+p}\frac{2}{1-\alpha_k}\right)(\|Tu_{3n+1} - u_{3n+1}\| - \|Tu_{3n+1+3p} - u_{3n+1+3p}\|).
\end{aligned} \tag{5}$$

And obviously

$$k \geq 3n-2,\ \|Tu_{k+1} - Tu_k\| \leq \beta_k\|Tu_{3n-2} - u_{3n-2}\|, \tag{6}$$

$$k > t,\ \|Tu_k - Tu_t\| \leq \|u_k - u_t\|. \tag{7}$$

Case 1: $p=3m$,m≥1. From (6) and (7)

$$\begin{aligned}
&\|Tu_{3n+1+p} - u_{3n+1}\| \\
&= \|Tx_{n+m+1} - x_{n+1}\| \\
&= \|Tx_{n+m+1} - Ty_n\| \\
&\le \|x_{n+m+1} - y_n\| \\
&= \|Ty_{n+m} - Tz_n\| \\
&\le \|y_{n+m} - z_n\| \\
&= \|Tz_{n+m} - (1-\alpha_n)x_n - \alpha_n Tx_n\|
\end{aligned}$$

$$\le (1-\alpha_n)\|Tz_{n+m} - x_n\| + \alpha_n\|Tz_{n+m} - Tx_n\|$$

$$\le (1-\alpha_n)\|Tz_{n+m} - x_n\|$$

$$+\alpha_n\left(\|Tz_{n+m} - Tx_{n+m}\| + \|Tx_{n+m} - Ty_{n+m-1}\| + \cdots + \|Ty_n - Tz_n\| + \|Tz_n - Tx_n\|\right)$$

$$= (1-\alpha_n)\|Tu_{3n-1+p} - u_{3n-2}\| + \alpha_n \sum_{k=3n-2}^{3n-2+p} \|Tu_{k+1} - Tu_k\|$$

$$\le (1-\alpha_n)\|Tu_{3n-1+p} - u_{3n-2}\| + \alpha_n \sum_{k=3n-2}^{3n-2+p} \beta_k \|Tu_{3n-2} - u_{3n-2}\|.$$

It follows that

$$\|Tu_{3n+1+p} - u_{3n+1}\| \le (1-\alpha_n)\|Tu_{3n-1+p} - u_{3n-2}\| + \alpha_n \sum_{k=3n-2}^{3n-2+p} \beta_k \|Tu_{3n-2} - u_{3n-2}\| \tag{8}$$

Using (5) and (8), we have

$$\left(1 + \sum_{k=3n+1}^{3n+p} \beta_k\right)\|Tu_{3n+1} - u_{3n+1}\|$$

$$\le (1-\alpha_n)\|Tu_{3n-1+p} - u_{3n-2}\| + \alpha_n \sum_{k=3n-2}^{3n-2+p} \beta_k \|Tu_{3n-2} - u_{3n-2}\|$$

$$+\left(\prod_{k=n+1}^{n+p} \frac{2}{1-\alpha_k}\right)\left(\|Tu_{3n+1} - u_{3n+1}\| - \|Tu_{3n+1+3p} - u_{3n+1+3p}\|\right).$$

From $\left(1 + \sum_{k=3n+1}^{3n+p} \beta_k\right) \le \left(\prod_{k=n+1}^{n+p} \frac{1}{1-\alpha_k}\right)$ and $\|Tu_{3n+1} - u_{3n+1}\| \le \|Tu_{3n-2} - u_{3n-2}\|$, we have

$$\left(1+\sum_{k=3n+1}^{3n+p}\beta_k\right)\|Tu_{3n-2}-u_{3n-2}\|$$
$$\leq(1-\alpha_n)\|Tu_{3n-1+p}-u_{3n-2}\|+\alpha_n\sum_{k=3n-2}^{3n-2+p}\beta_k\|Tu_{3n-2}-u_{3n-2}\|$$
$$+\left(\prod_{k=n+1}^{n+p}\frac{2}{1-\alpha_k}\right)(\|Tu_{3n-2}-u_{3n-2}\|-\|Tu_{3n+1+3p}-u_{3n+1+3p}\|).$$

Then

$$\frac{\left(1+\sum_{k=3n+1}^{3n+p}\beta_k-\alpha_n\sum_{k=3n-2}^{3n-2+p}\beta_k\right)}{1-\alpha_n}\|Tu_{3n-2}-u_{3n-2}\|$$
$$\leq\|Tu_{3n-1+p}-u_{3n-2}\|$$
$$+\left(\prod_{k=n}^{n+p}\frac{2}{1-\alpha_k}\right)(\|Tu_{3n-2}-u_{3n-2}\|-\|Tu_{3n+1+3p}-u_{3n+1+3p}\|).$$

It follows that

$$\left(1+\sum_{k=3n-2}^{3n-2+p}\beta_k\right)\|Tu_{3n-2}-u_{3n-2}\|$$
$$\leq\|Tu_{3n-1+p}-u_{3n-2}\|$$
$$+\left(\prod_{k=n}^{n+p}\frac{2}{1-\alpha_k}\right)(\|Tu_{3n-2}-u_{3n-2}\|-\|Tu_{3n+1+3p}-u_{3n+1+3p}\|).$$

Thus, for n, p+1, (4) holds.

Case 2: p=3m+1, m≥0. From (6) and (7), we have

$$\|Tu_{3n+1+p}-u_{3n+1}\|$$
$$=\|Tz_{n+m+1}-x_{n+1}\|$$
$$=\|Tz_{n+m+1}-Ty_n\|$$
$$\leq\|z_{n+m+1}-y_n\|$$
$$=\|(1-\alpha_{m+n+1})x_{m+n+1}+\alpha_{m+n+1}Tx_{m+n+1}-Tz_n\|$$

$$\leq (1-\alpha_{m+n+1})\|x_{m+n+1}-Tz_n\|+\alpha_{m+n+1}\|Tx_{m+n+1}-Tz_n\|$$
$$\leq (1-\alpha_{m+n+1})\|Ty_{m+n}-Tz_n\|+\alpha_{m+n+1}\|x_{m+n+1}-z_n\|$$
$$\leq (1-\alpha_{m+n+1})(\|Ty_{m+n}-Tz_{m+n}\|+\|Tz_{m+n}-Tx_{n+m}\|+\cdots+\|Tx_{n+1}-Ty_n\|$$
$$+\|Ty_n-Tz_n\|)+\alpha_{m+n+1}\|x_{m+n+1}-z_n\|$$
$$=(1-\alpha_{m+n+1})\sum_{k=3n-1}^{3n-2+p}\|Tu_{k+1}-Tu_k\|+\alpha_{m+n+1}\|Ty_{m+n}-(1-\alpha_n)x_n-\alpha_nTx_n\|$$
$$\leq (1-\alpha_{m+n+1})\sum_{k=3n-1}^{3n-2+p}\beta_k\|Tu_{3n-2}-u_{3n-2}\|$$
$$+\alpha_{m+n+1}((1-\alpha_n)\|Ty_{m+n}-x_n\|+\alpha_n\|Ty_{m+n}-Tx_n\|)$$
$$\leq (1-\alpha_{m+n+1})\sum_{k=3n-1}^{3n-2+p}\beta_k\|Tu_{3n-2}-u_{3n-2}\|+\alpha_{m+n+1}(1-\alpha_n)\|Ty_{m+n}-x_n\|$$
$$+\alpha_{m+n+1}\alpha_n(\|Ty_{m+n}-Tz_{m+n}\|+\|Tz_{m+n}-Tx_{m+n}\|+\cdots+\|Ty_n-Tz_n\|+\|Tz_n-Tx_n\|)$$
$$=(1-\alpha_{m+n+1})\sum_{k=3n-1}^{3n-2+p}\beta_k\|Tu_{3n-2}-u_{3n-2}\|+\alpha_{m+n+1}(1-\alpha_n)\|Ty_{m+n}-x_n\|$$
$$+\alpha_{m+n+1}\alpha_n\sum_{k=3n-2}^{3n-2+p}\|Tu_{k+1}-Tu_k\|$$
$$\leq (1-\alpha_{m+n+1})\sum_{k=3n-1}^{3n-2+p}\beta_k\|Tu_{3n-2}-u_{3n-2}\|+\alpha_{m+n+1}(1-\alpha_n)\|Ty_{m+n}-x_n\|$$
$$+\alpha_{m+n+1}\alpha_n\sum_{k=3n-2}^{3n-2+p}\beta_k\|Tu_k-u_k\|$$
$$=(1-\alpha_{m+n+1})\sum_{k=3n-1}^{3n-2+p}\beta_k\|Tu_{3n-2}-u_{3n-2}\|+\alpha_{m+n+1}(1-\alpha_n)\|Tu_{3n-1+p}-u_{3n-2}\|$$
$$+\alpha_{m+n+1}\alpha_n\sum_{k=3n-2}^{3n-2+p}\beta_k\|Tu_k-u_k\|$$
$$=(1-\alpha_{m+n+1}+\alpha_{m+n+1}\alpha_n)\sum_{k=3n-2}^{3n-2+p}\beta_k\|Tu_{3n-2}-u_{3n-2}\|$$
$$-\alpha_n(1-\alpha_{m+n+1})\|Tu_{3n-2}-u_{3n-2}\|+\alpha_{m+n+1}(1-\alpha_n)\|Tu_{3n-1+p}-u_{3n-2}\|.$$

It follows that

$$\|Tu_{3n+1+p} - u_{3n+1}\|$$
$$\leq (1 - \alpha_{m+n+1} + \alpha_{m+n+1}\alpha_n) \sum_{k=3n-2}^{3n-2+p} \beta_k \|Tu_{3n-2} - u_{3n-2}\| \tag{9}$$
$$-\alpha_n(1 - \alpha_{m+n+1})\|Tu_{3n-2} - u_{3n-2}\| + \alpha_{m+n+1}(1 - \alpha_n)\|Tu_{3n-1+p} - u_{3n-2}\|$$

Using (5) and (9), we have

$$\left(1 + \sum_{k=3n+1}^{3n+p} \beta_k\right)\|Tu_{3n+1} - u_{3n+1}\|$$
$$\leq (1 - \alpha_{m+n+1} + \alpha_{m+n+1}\alpha_n) \sum_{k=3n-2}^{3n-2+p} \beta_k \|Tu_{3n-2} - u_{3n-2}\|$$
$$-\alpha_n(1 - \alpha_{m+n+1})\|Tu_{3n-2} - u_{3n-2}\| + \alpha_{m+n+1}(1 - \alpha_n)\|Tu_{3n-1+p} - u_{3n-2}\|$$
$$+\left(\prod_{k=n+1}^{n+p} \frac{2}{1-\alpha_k}\right)(\|Tu_{3n+1} - u_{3n+1}\| - \|Tu_{3n+1+3p} - u_{3n+1+3p}\|).$$

From $\left(1 + \sum_{k=3n+1}^{3n+p} \beta_k\right) \leq \left(\prod_{k=n+1}^{n+p} \frac{1}{1-\alpha_k}\right)$ and $\|Tu_{3n+1} - u_{3n+1}\| \leq \|Tu_{3n-2} - u_{3n-2}\|$,

we have

$$\left(1 + \sum_{k=3n+1}^{3n+p} \beta_k\right)\|Tu_{3n-2} - u_{3n-2}\|$$
$$\leq (1 - \alpha_{m+n+1} + \alpha_{m+n+1}\alpha_n) \sum_{k=3n-2}^{3n-2+p} \beta_k \|Tu_{3n-2} - u_{3n-2}\|$$
$$-\alpha_n(1 - \alpha_{m+n+1})\|Tu_{3n-2} - u_{3n-2}\| + \alpha_{m+n+1}(1 - \alpha_n)\|Tu_{3n-1+p} - u_{3n-2}\|$$
$$+\left(\prod_{k=n+1}^{n+p} \frac{2}{1-\alpha_k}\right)(\|Tu_{3n-2} - u_{3n-2}\| - \|Tu_{3n+1+3p} - u_{3n+1+3p}\|).$$

Then

$$\frac{\left(1 + \sum_{k=3n+1}^{3n+p} \beta_k + \alpha_n(1 - \alpha_{m+n+1}) - (1 - \alpha_{m+n+1} + \alpha_{m+n+1}\alpha_n) \sum_{k=3n-2}^{3n-2+p} \beta_k\right)}{\alpha_{m+n+1}(1 - \alpha_n)}$$
$$\|Tu_{3n-2} - u_{3n-2}\| \leq \|Tu_{3n-1+p} - u_{3n-2}\|$$
$$+\left(\prod_{k=n}^{n+p} \frac{2}{1-\alpha_k}\right)(\|Tu_{3n-2} - u_{3n-2}\| - \|Tu_{3n+1+3p} - u_{3n+1+3p}\|).$$

It follows that

$$\left(1+\sum_{k=3n-2}^{3n-2+p}\beta_k\right)\|Tu_{3n-2}-u_{3n-2}\|$$
$$\leq \|Tu_{3n-1+p}-u_{3n-2}\|$$
$$+\left(\prod_{k=n}^{n+p}\frac{2}{1-\alpha_k}\right)\left(\|Tu_{3n-2}-u_{3n-2}\|-\|Tu_{3n+1+3p}-u_{3n+1+3p}\|\right).$$

Thus, for *n*,*p*+1 holds.

Case 3: *p*=3m+2,m≥0. From (6) and (7), we have

$$\|Tu_{3n+1+p}-u_{3n+1}\|$$
$$=\|Ty_{n+m+1}-Ty_n\|$$
$$\leq \|y_{n+m+1}-y_n\|$$
$$=\|Tz_{n+m+1}-Tz_n\|$$
$$\leq \|z_{n+m+1}-z_n\|$$
$$\leq \|z_{n+m+1}-(1-\alpha_n)x_n-\alpha_n Tx_n\|$$
$$\leq (1-\alpha_n)\|z_{n+m+1}-x_n\|+\alpha_n\|z_{n+m+1}-Tx_n\|$$
$$=(1-\alpha_n)\|(1-\alpha_{n+m+1})x_{n+m+1}+\alpha_{n+m+1}Tx_{n+m+1}-x_n\|$$
$$+\alpha_n\|(1-\alpha_{n+m+1})x_{n+m+1}+\alpha_{n+m+1}Tx_{n+m+1}-Tx_n\|$$
$$\leq (1-\alpha_n)((1-\alpha_{m+n+1})\|x_{n+m+1}-x_n\|+\alpha_{m+n+1}\|Tx_{n+m+1}-x_n\|)$$
$$+\alpha_n((1-\alpha_{m+n+1})\|x_{n+m+1}-Tx_n\|+\alpha_{m+n+1}\|Tx_{n+m+1}-Tx_n\|)$$
$$\leq (1-\alpha_n)\alpha_{m+n+1}\|Tx_{n+m+1}-x_n\|$$
$$+((1-\alpha_{m+n+1})(1-\alpha_n)+\alpha_{m+n+1}\alpha_n)\|x_{n+m+1}-x_n\|$$
$$+\alpha_n(1-\alpha_{m+n+1})\|Ty_{n+m+1}-Tx_n\|$$
$$\leq (1-\alpha_n)\alpha_{m+n+1}\|Tx_{n+m+1}-x_n\|$$
$$+((1-\alpha_{m+n+1})(1-\alpha_n)+\alpha_{m+n+1}\alpha_n)(\|Ty_{n+m}-Tz_{n+m}\|+\cdots+\|Tz_n-Tx_n\|+\|Tx_n-x_n\|)$$
$$+\alpha_n(1-\alpha_{m+n+1})(\|Ty_{n+m+1}-Tz_{n+m+1}\|+\cdots+\|Ty_n-Tz_n\|+\|Tz_n-Tx_n\|)$$
$$\leq (1-\alpha_n)\alpha_{m+n+1}\|Tx_{n+m+1}-x_n\|$$

$$+((1-\alpha_{m+n+1})(1-\alpha_n)+\alpha_{m+n+1}\alpha_n)\left(\sum_{k=3n-2}^{3n-3+p}\|Tu_{k+1}-Tu_k\|+\|Tu_{3n-2}-x_{3n-2}\|\right)$$

$$+(1-\alpha_{m+n+1})\alpha_n\sum_{k=3n-2}^{3n-3+p}\|Tu_{k+1}-Tu_k\|$$

$$\leq(1-\alpha_n)\alpha_{m+n+1}\|Tx_{n+m+1}-x_n\|$$

$$+((1-\alpha_{m+n+1})(1-\alpha_n)+\alpha_{m+n+1}\alpha_n)\sum_{k=3n-2}^{3n-2+p}\beta_k\|Tx_n-x_n\|$$

$$+(1-\alpha_{m+n+1})\alpha_n\sum_{k=3n-2}^{3n-3+p}\beta_k\|Tx_n-x_n\|$$

$$\leq(1-\alpha_n)\alpha_{m+n+1}\|Tu_{3n-1+p}-u_{3n-2}\|$$

$$+\left((1-\alpha_{m+n+1}+\alpha_n\alpha_{m+n+1})\sum_{k=3n-2}^{3n-2+p}\beta_k-\alpha_n(1-\alpha_{m+n+1})\right)\|Tu_{3n-2}-u_{3n-2}\|.$$

It follows that

$$\|Tu_{3n+1+p}-u_{3n+1}\|$$
$$\leq(1-\alpha_n)\alpha_{m+n+1}\|Tu_{3n-1+p}-u_{3n-2}\|$$
$$+\left((1-\alpha_{m+n+1}+\alpha_n\alpha_{m+n+1})\sum_{k=3n-2}^{3n-2+p}\beta_k-\alpha_n(1-\alpha_{m+n+1})\right)\|Tu_{3n-2}-u_{3n-2}\| \tag{10}$$

Using (5) and (10), we have

$$\left(1+\sum_{k=3n+1}^{3n+p}\beta_k\right)\|Tu_{3n+1}-u_{3n+1}\|$$

$$\leq(1-\alpha_n)\alpha_{m+n+1}\|Tu_{3n-1+p}-u_{3n-2}\|$$

$$+\left((1-\alpha_{m+n+1}+\alpha_n\alpha_{m+n+1})\sum_{k=3n-2}^{3n-2+p}\beta_k-\alpha_n(1-\alpha_{m+n+1})\right)\|Tu_{3n-2}-u_{3n-2}\|$$

$$+\left(\prod_{k=n+1}^{n+p}\frac{2}{1-\alpha_k}\right)(\|Tu_{3n+1}-u_{3n+1}\|-\|Tu_{3n+1+3p}-u_{3n+1+3p}\|).$$

From $\left(1+\sum_{k=3n+1}^{3n+p}\beta_k\right)\leq\left(\prod_{k=n+1}^{n+p}\frac{2}{1-\alpha_k}\right)$ and $\|Tu_{3n+1}-u_{3n+1}\|\leq\|Tu_{3n-2}-u_{3n-2}\|$,

$$\left(1+\sum_{k=3n+1}^{3n+p}\beta_k\right)\|Tu_{3n-2}-u_{3n-2}\|$$
$$\leq(1-\alpha_n)\alpha_{m+n+1}\|Tu_{3n-1+p}-u_{3n-2}\|$$
$$+\left((1-\alpha_{m+n+1}+\alpha_n\alpha_{m+n+1})\sum_{k=3n-2}^{3n-2+p}\beta_k-\alpha_n(1-\alpha_{m+n+1})\right)\|Tu_{3n-2}-u_{3n-2}\|$$
$$+\left(\prod_{k=n+1}^{n+p}\frac{2}{1-\alpha_k}\right)(\|Tu_{3n-2}-u_{3n-2}\|-\|Tu_{3n+1+3p}-u_{3n+1+3p}\|).$$

Then

$$\frac{(1+\sum_{k=3n+1}^{3n+p}\beta_k-\left((1-\alpha_{m+n+1}+\alpha_n\alpha_{m+n+1})\sum_{k=3n-2}^{3n-2+p}\beta_k-\alpha_n(1-\alpha_{m+n+1})\right)}{(1-\alpha_n)\alpha_{m+n+1}}\|Tu_{3n-2}-u_{3n-2}\|$$
$$\leq\|Tu_{3n-1+p}-u_{3n-2}\|$$
$$+\left(\prod_{k=n}^{n+p}\frac{2}{1-\alpha_k}\right)(\|Tu_{3n-2}-u_{3n-2}\|-\|Tu_{3n+1+3p}-u_{3n+1+3p}\|).$$

It follows that

$$\left(1+\sum_{k=3n-2}^{3n-2+p}\beta_k\right)\|Tu_{3n-2}-u_{3n-2}\|$$
$$\leq\|Tu_{3n-1+p}-u_{3n-2}\|$$
$$+\left(\prod_{k=n}^{n+p}\frac{2}{1-\alpha_k}\right)(\|Tu_{3n-2}-u_{3n-2}\|-\|Tu_{3n+1+3p}-u_{3n+1+3p}\|).$$

Thus, for *n*, *p*+1, (4) holds. This completes the induction.

Lemma 3. $T:C\to C$ is a T-(D_a) mapping, $\|Tx-x\|\leq\|Ty-y\|$. Then

$$\|x-Ty\|\leq 3\|Tx-x\|+\|x-y\|.$$

Proof: Since $\|Tx-x\|\leq\|Ty-y\|$, we have $Tx\in C(T,y,\alpha)$. Then

$$\|T^2x-Ty\|\leq\|Tx-y\|.$$

It follows that

$$\|x-Ty\|\leq\|x-Tx\|+\|T^2x-Tx\|+\|T^2x-Ty\|.$$

From **Proposition 3**, we have

$$\|x-Ty\|\leq 2\|Tx-x\|+\|Tx-y\|\leq 2\|Tx-x\|+\|Tx-x\|+\|x-y\|=3\|Tx-x\|+\|x-y\|.$$

Theorem 1. Assume that $T{:}C\to C$ is a T-(D_a) mapping and $\{x_n\},\{y_n\},\{z_n\}$ are sequences generated by iteration (3), $\frac{1}{2} < a \le \alpha_n \le b < 1$. Then $\lim_{n\to\infty}\|Tx_n - x_n\| = 0$.

Proof: Since C is bounded, there must exists d>0, for every x∈C, ‖x‖≤d. Let $\{u_m\}$ satisfy $u_{3n-2}=x_n, u_{3n-1}=z_n, u_{3n}=y_n$. From Lemma 1, $\lim_{k\to\infty}\|Tu_k-u_k\|=r\ge 0$ Assume r>0. Let □ satisfy

$$e^{\frac{6}{1-b}\left(\frac{d}{r}+1\right)}\varepsilon < r$$

and choose n so that for every p>0

$$\|Tu_{3n-2} - u_{3n-2}\| - \|Tu_{3n-2+3p} - u_{3n-2+3p}\| < \varepsilon.$$

Now choose p so that $r\left(\sum_{k=3n-2}^{3n+p-4}\beta_k\right) \le d \le r\left(\sum_{k=3n-2}^{3n+p-3}\beta_k\right)$.

Since $\frac{1}{2} < a \le \alpha_n \le b < 1$, for every k,t, we have $1+\alpha_k<3\alpha_k, \alpha_t<2\alpha_k$. From **Lemma 2** and $r\le\|Tu_{3n-2}-u_{3n-2}\|$, we have

$$d + r \le r\left(1 + \sum_{k=3n-2}^{3n+p-3}\beta_k\right)$$

$$\le \left(1 + \sum_{k=3n-2}^{3n+p-3}\beta_k\right)\|Tu_{3n-2} - u_{3n-2}\|$$

$$\le \|Tu_{3n-2+p} - u_{3n-2}\|$$

$$+\left(\prod_{k=n}^{n+p-1}\frac{2}{1-\alpha_k}\right)\left(\|Tu_{3n-2} - u_{3n-2}\| - \|Tu_{3n-2+3p} - u_{3n-2+3p}\|\right)$$

$$< d + \left(\prod_{k=n}^{n+p-1}\frac{2}{1-\alpha_k}\right)\varepsilon$$

$$= d + e^{\sum_{k=n}^{n+p-1}\ln\left(1+\frac{1+\alpha_k}{1-\alpha_k}\right)}\varepsilon$$

$$\le d + e^{\sum_{k=n}^{n+p-1}\frac{1+\alpha_k}{1-\alpha_k}}\varepsilon$$

$$\le d + e^{\frac{3}{1-b}\sum_{k=n}^{n+p-1}\alpha_k}\varepsilon$$

$$\le d + e^{\frac{6}{1-b}\sum_{k=3n-2}^{3n+p-3}\beta_k}\varepsilon$$

$$\leq d + e^{\frac{6}{1-b}\left(\sum_{k=3n-2}^{3n+p-4}\beta_k + 1\right)}\varepsilon$$

$$< d + e^{\frac{6}{1-b}\left(\frac{d}{r}+1\right)}\varepsilon < d + r.$$

This is a contradiction. So $\lim_{k\to\infty}\|Tu_k - u_k\| = 0$. That is to say, $\lim_{n\to\infty}\|Tx_n - x_n\| = 0$. This completes the proof.

Theorem 2. Assume that $T:C\to C$ is a T-(D_a) mapping and $\{x_n\}$ is generated by iteration (3), $\frac{1}{2} < a \leq \alpha_n \leq b < 1$. Then the sequence $\{x_n\}$ converges to a fixed point of T.

Proof: Since C is compact, there exists a subsequence $\{x_{nk}\} \subset \{x_n\}$ which converges to some $z \in C$. By **Lemma 3**, we have $\|x_{nk} - T_z\| \leq 3\|Tx_{nk} - x_{nk}\| + \|x_{nk} - z\|$. Since $\lim_{nk\to\infty}\|Tx_{nk} - x_{nk}\| = 0$ and $\lim_{nk\to\infty}\|x_{nk} - z\| = 0$, we have $\lim_{nk\to\infty}\|x_{nk} - T_z\| = 0$. This implies that z=Tz. On the other hand, from **Proposition 3**

$$\begin{aligned}\|x_{n+1} - z\| &\leq \|y_n - z\| \leq \|z_n - z\| \\ &\leq \alpha_n\|Tx_n - z\| + (1-\alpha_n)\|x_n - z\| \\ &\leq \|x_n - z\|.\end{aligned}$$

So, $\lim_{n\to\infty}\|x_n - z\|$ exists. Therefore, $\lim_{n\to\infty}\|x_n - z\| = 0$. This completes the proof.

AT A GLANCE

This work presents a T – (Da) mapping that is less strong than the nonexpansive mapping. It presents multiple iterations for determining the fixed point of the T – (Da) mapping. This provides fixed point theorems and convergence theorems for the T – (Da) mapping in a Banach space, rather than a uniformly convex Banach space. This paper presents fundamental features of the T – (Da) mapping and provides an illustrative example to demonstrate the existence of the T – (Da) mapping. The findings of this study are derived in a broad range of Banach spaces. This paper examines the necessary conditions for the convergence of fixed points of mappings in general Banach spaces under greater iterations.

REFERENCES

1. Suzuki T. Fixed point theorems and convergence theorems for some generalized nonexpansive mappings. Journal of Mathematical Analysis and Applications. 2008;340:1088-1095
2. Thakur BS, Thakur D, Postolache M. A new iterative scheme for numerical reckoning fixed points of Suzuki's generalized nonexpansive mappings. Journal of Applied Mathematics and Computing. 2016;275:147-155

04 FOLDING ON THE CHAOTIC GRAPH OPERATIONS AND THEIR FUNDAMENTAL GROUP

1. INTRODUCTION: AN OVERVIEW

In recent decades, there has been increased focus on studying social, biological, and communication networks. Graphical representations of these networks and systems have shown to be quite useful. These representations are commonly used to verify or illustrate the connections or linkages between different components of such networks [1, 2].

A graph is an ordered $G = (V(G), E(G))$ where $V(G) \neq \varphi$, $E(G)$ is a set disjoint from $V(G)$, elements of $V(G)$ are called the vertices of G, and elements of $E(G)$ are called the edges. The foundation stone of graph theory was laid by Euler in 1736 by solving a puzzle called Königsberg seven-bridge problem as in Figure 1[1, 3].

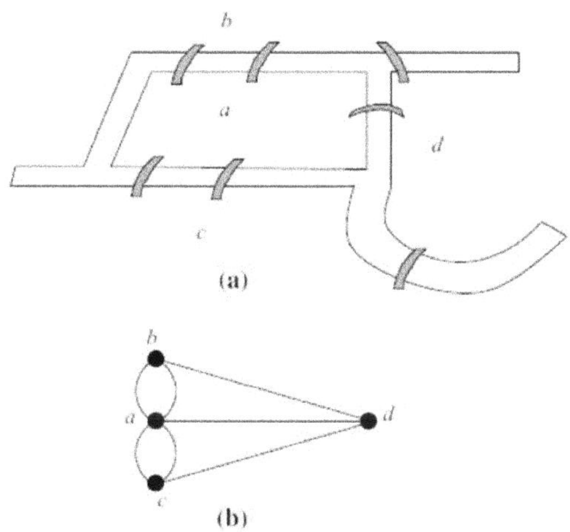

Figure 1. Königsberg seven-bridge problem.

There are many graphs with which one can construct a new graph from a given graph or set of graphs, such as the Cartesian product and the line graph. A graph G is a finite non-empty set V of objects called vertices (the singular is vertex) together with a set E of two-element subsets of V called edges. The number of vertices in a graph G is the order of G, and the number of edges is the size of G. To indicate that a graph G has vertex set V and edge set E, we sometimes write G = (V, E). To emphasize that V is the vertex set of a graph G, we often write V as V(G). For the same reason, we also write E as E(G). A graph H is said to be a subgraph of a graph G if $V(H) \subseteq V(G)$ and $E(H) \subseteq E(G)$. The complete graph with n-vertices will be denoted by K_n. A null graph is a graph containing no edges; the null graph with n-vertices is denoted by N_n. A cycle graph is a graph consisting of a single cycle, the cycle graph with n-vertices is denoted by C_n. The path graph is a graph consisting of a single path; the path graph with n-vertices is denoted by P_n [1, 2, 3, 4, 5, 6, 7, 8, 9, 10, 11]. Let G and H be two graphs. A function $\varphi:V(G)\to V(H)$ is a homomorphism from G to H if it preserves edges, that is, if for every edge $e\in E(G), f(e)\in E(H)$ [12, 13]. A core is a graph which does not retract to a proper subgraph. Any graph is homomorphically equivalent to a unique core [7].

The folding is a continuous function $f:G\to H$ such that for each $v\in V(G), f(v)\in V(H)$, and for each $e\in E(G), f(e)\in E(H)$ [14]. Let X be a space, and let I be the unit interval [0,1] in R, a homotopy of paths in X is a family $g_t:I\to X, 0\leq t\leq 1$ such that (i) the endpoints $g_t(0)=x_0$ and $g_t(1)=x_1$ are independent of t and (ii) the associated map $G:I\times I\to X$ defined by $G(s,t) = g_t(s)$ is continuous [15]. Given spaces X and Y with chosen points $x_0\in X$, and $y_0 \in Y$, the wedge sum $X\vee Y$ is the quotient of the disjoint union $X\cup Y$ obtained identifying x_0 and y_0 to a single point [15]. Two spaces X and Y are of the "same homotopy type" if there exist continuous maps $f:X\to Y$ and $g:Y\to X$ such that $g\circ f\cong I_X$: $X\to X$ and $f\circ g\cong I_Y$: $Y\to Y$ [16]. The fundamental group briefly consists of equivalence classes of homotopic closed paths with the law of composition following one path to another. However, the set of homotopy classes of loops based at the point x_0 with the product operation $[f][g]=[f\cdot g]$ is called the fundamental group and denoted by $\pi_1(X,x_0)$ [4, 17, 18, 19, 20, 21, 22, 23, 24]. Over many years, chaos has been shown to be an interesting and even common phenomenon in nature. Chaos has been shown to exist in a wide variety of settings: in fluid dynamics such as Raleigh-Bernard convection, in chemistry such as the Belousov-Zhabotinsky reaction, in nonlinear optics in certain lasers, in celestial mechanics, in electronics in the flutter of an overdriven airplane wing, some models of population dynamics, and likely in meteorology, physiological oscillations such as certain heart rhythms, as well as brain patterns [17, 24, 25, 26, 27, 28, 29, 30]. AI algorithms related to adjacency matrices on the operations of the graph are discussed in [31, 32].

2. THE MAIN RESULTS

First, we will introduce the following:

Definition 1. *The chaotic edge* $\bar{e}$ *is a geometric edge* e_1 *that carries many other edges* $(e_2,e_3,...)$ *each one of them homotopic to the original one as in* Figure 2. *Also the* *chaotic vertices of* $\overline{e}$ *are* $\overline{v} = (v_1, v_2, \ldots)$ *and* $\overline{u} = (u_1, u_2, \ldots)$. *For chaotic edge* $\overline{e}$, *we have two cases:*

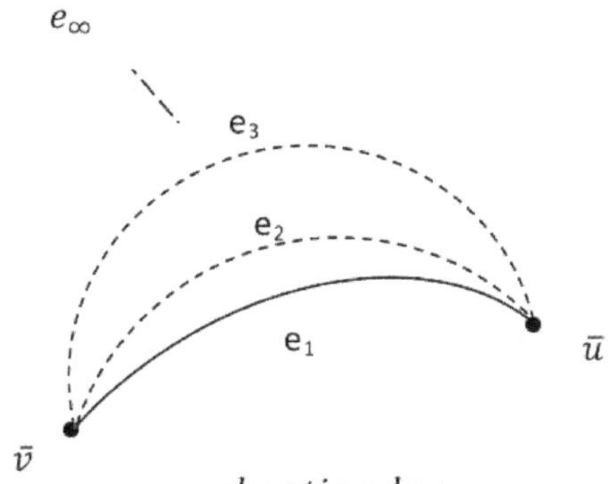

chaotic edge

Figure 2. Chaotic edge.

Case 1 (1) *e_1, e_2, e_3,... are of the same physical properties.*

Case 2 (2) *e_1, e_2, e_3,... represent different physical properties; for example, e_1 represents density, e_2 represents hardness, e_3 represents magnetic fields, and so on.*

Definition 2. *A chaotic graph* $\overline{G}$ *is a collection of finite non-empty set* $\overline{V}$ *of objects called chaotic vertices together with a set* $\overline{E}$ *of two-element subsets of* $\overline{V}$ *called chaotic edges. The number of chaotic edges is the size of* $\overline{G}$.

Definition 3. *Given chaotic connected graphs* $\overline{G}_1$ *and* $\overline{G}_2$ *with given edges* $\overline{e}_1 \in \overline{G}_1$ *and* $\overline{e}_2 \in \overline{G}_2$, *then the chaotic connected edge graph* $\overline{G}_1\ \overline{G}_2$ *is the quotient of disjoint union* $\overline{G}_1 \cup \overline{G}_2$ *acquired by identifying two chaotic edges* $\overline{e}_1$ *and* $\overline{e}_2$ *to a single chaotic edge (up to chaotic isomorphism) as in* Figure 3.

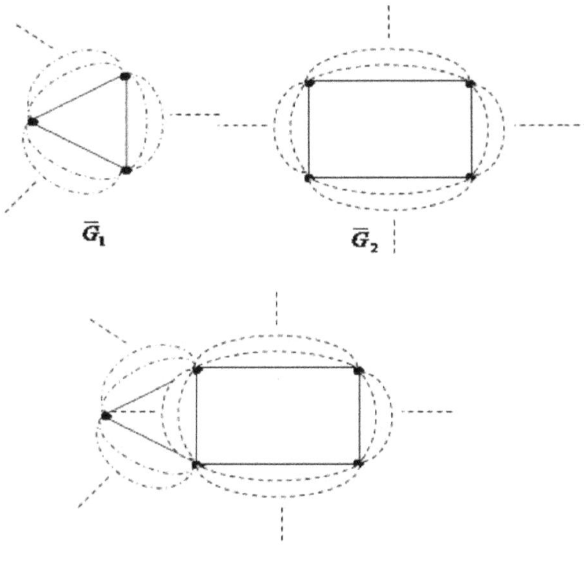

$\bar{G}_1 \veebar \bar{G}_2$

Figure 3. Chaotic connected edge.

Definition 4. *A chaotic graph* $(\overline{H})$ *is called a chaotic subgraph of a chaotic graph* $\overline{G}$ *if* $\overline{V}(\overline{H}) \subseteq \overline{V}(\overline{G})$ *and* $\overline{E}(\overline{H}) \subseteq \overline{E}(\overline{G})$.

Definition 5. *Let* $\overline{G}$ *and* $(\overline{H})$ *be two chaotic graphs. A function* $\overline{\varphi}: \overline{V}(\overline{G}) \to \overline{V}(\overline{H})$ *is chaotic homomorphism from* $\overline{G}$ *to* $(\overline{H})$ *if it preserves chaotic edges, that is, if for any chaotic edge* $[\overline{u}, \overline{v}]$ *of* $[\overline{\varphi}(\overline{u}), \overline{\varphi}(\overline{v})]$ *is a chaotic edge of* $(\overline{H})$

Definition 6. *A chaotic folding of a graph* $\overline{G}$ *is a chaotic subgraph* $(\overline{H})$ *of* $\overline{G}$ *such that there exists a chaotic homomorphism* $\overline{f}: \overline{G} \to (\overline{H})$, *called chaotic folding with* $\overline{f}(\overline{x}) = \overline{x}$ *for every chaotic vertex* $\overline{x}$ *of* $\overline{H}$.

Definition 7. *A chaotic core is a chaotic graph which does not chaotic retract to chaotic proper subgraph.*

Theorem 1. *Let* $\overline{G}_1$ *and* $\overline{G}_2$ *be two chaotic connected graphs. Then* $\overline{\pi}_1(\overline{G}_1 \veebar \overline{G}_2) = \overline{\pi}_1(\overline{G}_1) * \overline{\pi}_1(\overline{G}_2)$.

Proof. Let $\overline{G}_1$ and $\overline{G}_2$ be two chaotic connected graphs. Since $\overline{G}_1 \veebar \overline{G}_2$ and $\overline{G}_1 \vee \overline{G}_2$ are of same chaotic homotopy type, it follows that $\overline{\pi}_1(\overline{G}_1 \veebar \overline{G}_2) \approx \overline{\pi}_1(\overline{G}_1) * \overline{\pi}_1(\overline{G}_2)$. Hence, $\overline{\pi}_1(\overline{G}_1 \veebar \overline{G}_2) = \overline{\pi}_1(\overline{G}_1) * \overline{\pi}_1(\overline{G}_2)$.

Theorem 2. *The chaotic graphs* $\overline{G}_1$ *and* $\overline{G}_2$ *are chaotic subgraphs of* $\overline{G}_1 \veebar \overline{G}_2$. *Also, for any chaotic tree* $\overline{G}_1$ *and* $\overline{G}_2$, $\overline{G}_1 \veebar \overline{G}_2$ *is also chaotic tree and* $\overline{\pi}_1(\overline{G}_1 \veebar \overline{G}_2) = \overline{0}$.

Proof. The proof of this theorem is clear.

Theorem 3. *If* $\overline{G}_1, \overline{G}_2, \ldots, \overline{G}_n$ *are connected graphs, and* $\langle \overline{f}_1, \overline{f}_2, \ldots, \overline{f}_n \rangle$ *is a sequence of chaotic topological foldings of* $\veebar_{i=1}^{n} \overline{G}_i$ *into itself, then there is an induced sequence* $\langle \overline{\overline{f}}_1, \overline{\overline{f}}_2, \ldots, \overline{\overline{f}}_n \rangle$ *of non-trivial chaotic topological folding* $\overline{\overline{f}}_j : *_{i=1}^{n} \overline{\pi}_1(\overline{G}_{ii}) \to *_{i=1}^{n} \overline{\pi}_1(\overline{G}_{ii})$, *j=1,2,..., such that* $\overline{\overline{f}}_j(*_{i=1}^{n} \overline{\pi}_1(\overline{G}_{ii}))$ *reduces the rank of* $*_{i=1}^{n} \overline{\pi}_1(\overline{G}_{ii})$.

Proof. Consider the following sequence of topological foldings $\langle \overline{f}_1, \overline{f}_2, \ldots, \overline{f}_n \rangle$, where $\overline{f}_1 : \veebar_{i=1}^{n} \overline{G}_i \to \veebar_{i=1}^{n} \overline{G}_i$, is a topological foldings from $\vee_{i=1}^{n} \overline{G}_i$ into itself such that $\overline{f}_1(\veebar_{i=1}^{n} \overline{G}_i) = \overline{G}_1 \veebar \overline{G}_2 \veebar \ldots \veebar \overline{f}_1(\overline{G}_s) \veebar \ldots \veebar$ for s=1,2, ...n.

Since size $(\overline{f}_1(\overline{G}_s)) \leq size(\overline{G}_s)$ and $\overline{\overline{f}}_1(\overline{\pi}_1(\overline{G}_i)) = \overline{\pi}_1(\overline{f}_1(\overline{G}_i))$, it follows that rank $(\overline{\overline{f}}_1(\overline{\pi}_1(\overline{G}_s))) = \text{rank}(\overline{\pi}_1(\overline{f}_1(\overline{G}_s))) \leq \text{rank}(\overline{\pi}_1(\overline{G}_s))$, and so $\overline{\overline{f}}_1$ reduces the rank of $*_{i=1}^{n} \overline{\pi}_1(\overline{G}_{ii})$. Also if $\overline{f}_2(\veebar_{i=1}^{n} \overline{G}_i) = \overline{G}_1 \veebar \overline{G}_2 \veebar \ldots \veebar \overline{f}_2(\overline{G}_s) \veebar \ldots \veebar \overline{f}_2(\overline{G}_k) \veebar \ldots \veebar \overline{G}_n$ *for* *k*=1,2,...*n* and *s*<*k* and *size* $(\overline{f}_2(\overline{G}_s)) \leq size(\overline{G}_s)$ and size $(\overline{f}_2(\overline{G}_k)) \leq size(\overline{G}_k)$, we haverank $(\overline{\overline{f}}_2(\overline{\pi}_1(\overline{G}_s))) = \text{rank}(\overline{\pi}_1(\overline{f}_2(\overline{G}_s))) \leq \text{rank}(\overline{\pi}_1(\overline{G}_s))$, $\text{rank}(\overline{\overline{f}}_2(\overline{\pi}_1(\overline{G}_k))) = \text{rank}(\overline{\pi}_1(\overline{f}_2(\overline{G}_k))) \leq \text{rank}(\overline{\pi}_1(\overline{G}_k))$; thus $\overline{\overline{f}}_2$ reduces the rank of $*_{i=1}^{n} \overline{\pi}_1(\overline{G}_{ii})$. Moreover, by continuing with this procedure if $\overline{f}_n(\veebar_{i=1}^{n} \overline{G}_i) = \veebar_{i=1}^{n}(\overline{f}_n(\overline{G}_i))$, then $\overline{f}_n(*_{i=1}^{n} \overline{\pi}_1(\overline{G}_{ii})) = \overline{\pi}_1(\overline{f}_n(\veebar_{i=1}^{n} \overline{G}_i)) = \overline{\pi}_1(\veebar_{i=1}^{n} \overline{f}_n(\overline{G}_i)) \approx *_{i=1}^{n} \overline{\pi}_1(\overline{f}_n(\overline{G}_{ii}))$. Hence, $\overline{\overline{f}}_n$ reduces the rank of $*_{i=1}^{n} \overline{\pi}_1(\overline{G}_{ii})$.

Theorem 4. *Let $\overline{G}_1$ and $\overline{G}_2$ be two chaotic graphs; then there is a chaotic homomorphism $\overline{f}:\overline{G}_1\to\overline{G}_2$ which induces $\overline{\overline{f}}:\overline{\pi}_1(\overline{G}_1)\to\overline{\pi}_1(\overline{G}_2)$ if $\overline{\pi}_1(\overline{G}_2)$ is a chaotic folding* $\overline{\pi}_1(\overline{G}_1\veebar\overline{G}_2)$.

Proof. Let $\overline{f}:\overline{G}_1\to\overline{G}_2$ be a chaotic homomorphism. Since $\overline{G}_2$ is chaotic subgraph of $\overline{G}_1\veebar\overline{G}_2$, then there exists a chaotic homomorphism $\overline{f}:\overline{G}_1\veebar\overline{G}_2\to\overline{G}_2$ with $\overline{f}(\overline{v})=\overline{v}$ for any chaotic vertex $\overline{v}$ of $\overline{G}_2$ which induces $\overline{\overline{f}}:\overline{\pi}_1(\overline{G}_1)\to\overline{\pi}_1(\overline{G}_2)$. What follows from $\overline{G}_2$ is a chaotic folding of $\overline{G}_1\veebar\overline{G}_2$ in that $\overline{\pi}_1(\overline{G}_2)$ is a chaotic folding of $\overline{\pi}_1(\overline{G}_1\veebar\overline{G}_2)$. Conversely, assume that $\overline{G}_2$ is a chaotic folding of $\overline{G}_1\veebar\overline{G}_2$; thus $\overline{f}:\overline{G}_1\veebar\overline{G}_2\to\overline{G}_2$ is a chaotic homomorphism with $\overline{f}(\overline{v})=\overline{v}$ for any chaotic vertex $\overline{v}$ of $\overline{G}_2$, and so there is a chaotic homomorphism $\overline{f}:\overline{G}_1\to\overline{G}_2$ which induce $\overline{\overline{f}}:\overline{\pi}_1(\overline{G}_1)\to\overline{\pi}_1(\overline{G}_2)$.

Theorem 5. *For any chaotic path graphs $\overline{P}_n,\overline{P}_m, n, m\geq 2,$, there is a sequence of topological foldings with variation curvature $\{\overline{f}_i: i=1,2,\dots k\}$ on $\overline{P}_n\veebar\overline{P}_m$ which induce a sequence of topological foldings $\{\overline{\overline{f}}_i: i=1,2,\dots k\}$ such that $\overline{\overline{f}}_k(\overline{\overline{f}}_{k-1}(\dots(\overline{\overline{f}}_1(\overline{\pi}_1(\overline{P}_n\veebar\overline{P}_m)\dots\}=\overline{Z}$* and $lim_{k\to\infty}(\overline{\overline{f}}_k(\overline{\overline{f}}_{k-1}(\dots(\overline{\overline{f}}_1(\overline{\pi}_1(\overline{P}_n\veebar\overline{P}_m)\dots))=\overline{0}$.

Proof. Consider the following sequence of chaotic topological foldings with variation curvature, $\overline{f}_1:\overline{P}_n\veebar\overline{P}_m\to(\overline{P}_n\veebar\overline{P}_m)_1$, where $(\overline{P}_n\veebar\overline{P}_m)_1$ is a chaotic subgraph with decreasing inner curvature between every two adjacent chaotic edges in $\overline{P}_n\veebar\overline{P}_m$ and $\overline{f}_2:\overline{f}_1(\overline{P}_n\veebar\overline{P})_m\to\overline{f}_1((\overline{P}_n\veebar\overline{P}_m)_1)$ where $\overline{f}_2(\overline{f}_1((\overline{P}_n\veebar\overline{P}_m)_1))$ is a chaotic subgraph with decreasing inner curvature between every two adjacent chaotic edges in $\overline{f}_1((\overline{P}_n\veebar\overline{P}_m)_1)$, and so on, such that $\overline{f}_k(\overline{f}_{k-1}(\overline{f}_{k-2}(\dots(\overline{f}_1(\overline{P}_n\veebar\overline{P}_m)\dots)=\overline{C}_{n+m-2}$ and $\lim_{k\to\infty}(\overline{f}_k(\overline{f}_{k-1}(\overline{f}_{k-2}(\dots(\overline{f}_1(\overline{P}_n\veebar\overline{P}_m)\dots)=\overline{N}_1$, thus $\overline{\overline{f}}_k(\overline{\overline{f}}_{k-1}(\overline{\overline{f}}_{k-2}(\dots(\overline{\overline{f}}_1(\overline{\pi}_1(\overline{P}_n\veebar\overline{P}_m)\dots)=\overline{\pi}_{11}(\overline{C}_{n+m-2})=\overline{Z}$. Also, $\lim_{k\to\infty}(\overline{\overline{f}}_k(\overline{\overline{f}}_{k-1}(\overline{\overline{f}}_{k-2}(\dots(\overline{\overline{f}}_1(\overline{\pi}_1\ (\overline{P}_n\veebar\overline{P}_m)\dots)=\overline{\pi}_1(\overline{N}_1)=\overline{0}$.

Theorem 6. *For every two chaotic connected graphs $\overline{G}_1$ and $\overline{G}_2$, the fundamental group of the limit of chaotic topological folding of $\overline{G}_1\veebar\overline{G}_2=\overline{0}$.*

Proof. Let $\overline{G}_1$ and $\overline{G}_2$ be two chaotic connected graphs; then we have two cases:

Case (1): If $\overline{f}_1:\overline{G}_1\veebar\overline{G}_2\to\overline{G}_1\veebar\overline{G}_2$ is a chaotic topological folding such that $\overline{f}_1(\overline{G}_1\veebar\overline{G}_2)$ consists of chaotic cycles, so we can define a sequence of chaotic topological folding $\overline{f}_2:\overline{f}_1(\overline{G}_1\veebar\overline{G}_2)\to\overline{f}_1(\overline{G}_1\veebar\overline{G}_2)$ where $\overline{f}_2(\overline{f}_1(\overline{G}_1\veebar\overline{G}_2))$ is a chaotic tree with $\overline{n}$ chaotic edges, $\overline{f}_3:\overline{f}_2(\overline{f}_1(\overline{G}_1\veebar\overline{G}_2))\to\overline{f}_2(\overline{f}_1(\overline{G}_1\veebar\overline{G}_2))$, such that $\overline{f}_3(\overline{f}_2(\overline{f}_1(\overline{G}_1\veebar\overline{G}_2)\dots)$ is a chaotictreewith $\overline{k}<\overline{n}$ chaoticedges,chaoticedgesbycontinuingthisprocessweget $\overline{f}_k:\overline{f}_{k-1}(\overline{f}_{k-2}(\dots(\overline{f}_1(\overline{G}_1\veebar\overline{G}_2)\dots)\to\overline{f}_{k-1}(\overline{f}_{k-2}(\dots(\overline{f}_1(\overline{G}_1\veebar\overline{G}_2)\dots)$ such that $\lim_{k\to\infty}(\overline{f}_k(\overline{f}_{k-1}(\overline{f}_{k-2}(\dots(\overline{f}_1(\overline{G}_1\veebar\overline{G}_2)\dots)$ is a chaotic edge, and so $\overline{\pi}_1(\lim_{k\to\infty}(\overline{f}_k(\overline{f}_{k-1}(\overline{f}_{k-2}(\dots(\overline{f}_1(\overline{G}_1\veebar\overline{G}_2)\dots))=\overline{0}$.

Case (2): If $\overline{g}_1:\overline{G}_1\veebar\overline{G}_2\to\overline{G}_1\veebar\overline{G}_2$ is a chaotic topological folding such that.

$\overline{g}_1(\overline{G}_1\veebar\overline{G}_2)$ has no chaotic cycles, then clearly $\lim_{k\to\infty}(\overline{g}_k(\overline{g}_{k-1}(\overline{g}_{k-2}(\dots(\overline{g}_1(\overline{G}_1\veebar\overline{G}_2)\dots)$ is a chaotic edge and $\overline{\pi}_1(\lim_{k\to\infty}(\overline{g}_k(\overline{g}_{k-1}(\overline{g}_{k-2}(\dots(\overline{g}_1(\overline{G}_1\veebar\overline{G}_2)\dots))=\overline{0}$.

Theorem 7. *If $\overline{G}_1$ and $\overline{G}_2$ are chaotic connected and not chaotic cores graphs, then* $\overline{\pi}_1\left(\lim_{n\to\infty}\overline{f}_n(\overline{G}_1 \veebar \overline{G}_2)\right) = \overline{\pi}_1\left(\lim_{n\to\infty}\overline{f}_n(\overline{G}_1)\right) * \overline{\pi}_1\left(\lim_{n\to\infty}\overline{f}_n(\overline{G}_2)\right)$.

Proof. If $\overline{G}_1$ and $\overline{G}_2$ are chaotic connected and not chaotic cores graphs, then we get the following chaotic induced graphs $\lim_{n\to\infty}\overline{f}_n(\overline{G}_1 \veebar \overline{G}_2)$, $\lim_{n\to\infty}\overline{f}_n(\overline{G}_1)$, $\lim_{n\to\infty}\overline{f}_n(\overline{G}_2)$, and each of them are isomorphic to $\overline{k}_2$. Since $\overline{k}_2 \approx \overline{k}_2 \veebar \overline{k}_2$ it follows that $\lim_{n\to\infty}\overline{f}_n(\overline{G}_1 \veebar \overline{G}_2) = \lim_{n\to\infty}\overline{f}_n(\overline{G}_1) \veebar \lim_{n\to\infty}\overline{f}_n(\overline{G}_2)$ and $\overline{\pi}_1\left(\lim_{n\to\infty}\overline{f}_n(\overline{G}_1 \veebar \overline{G}_2)\right) = \overline{\pi}_1\left(\lim_{n\to\infty}\overline{f}_n(\overline{G}_1)\right) * \overline{\pi}_1\left(\lim_{n\to\infty}\overline{f}_n(\overline{G}_2)\right)$.

3. SOME APPLICATIONS

- A polymer is composed of many repeating units called monomers. Starch, cellulose, and proteins are natural polymers. Nylon and polyethylene are synthetic polymers. Polymerization is the process of joining monomers. Polymers may be formed by addition polymerization; furthermore, one essential advance likewise polymerization is mix as in Figure 4, which happens when the polymer's development is halted by free electrons from two developing chains that join and frame a solitary chain. The accompanying chart portrays mix, with the image (R) speaking to whatever remains of the chain.
- Chemical nature of enzymes, all known catalysts are proteins. They are high atomic weight mixes made up primarily of chains of amino acids connected together by peptide bonds as in Figure 5.
- There are two types of the subunit structure of ribosomes as in Figure 6 which is represented by the different connected types of protein subunit and rRNA to form a new type of ribosomes.

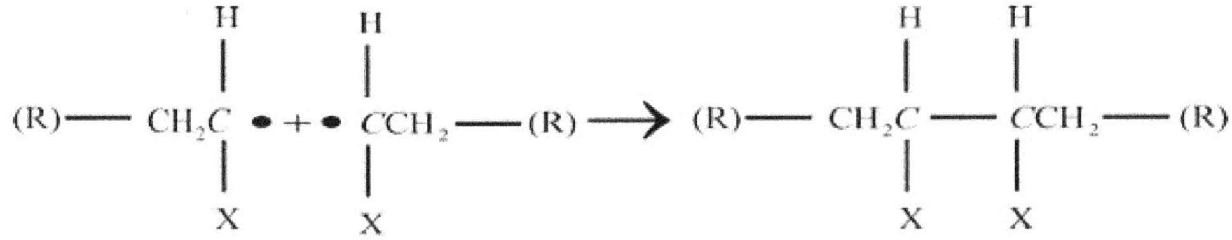

Figure 4. Polymerization.

4. CONCLUSION

In this chapter, the fundamental group of the limit chaotic foldings on chaotic connected edge graphs is deduced. Also, we can deduce some algorithms from a new operation of a graph by using the adjacency matrices.

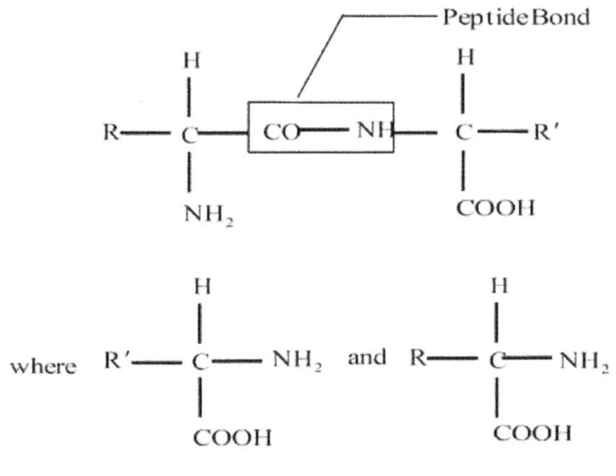

represent two typicalamino acids

Figure 5. Typical amino acids.

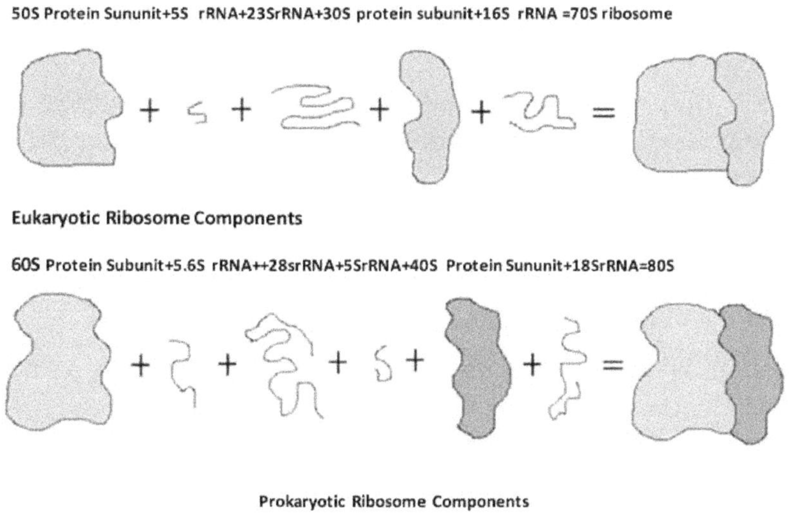

Figure 6. Prokaryotic ribosome components.

AT A GLANCE

The aim of this chapter is to introduce a new series of operations on the chaotic graph, specifically chaotic connected edge graphs, by utilizing the identification topology. This discussion will examine the concept of chaotic foldings on the chaotic edge graph, with a specific focus on the viewpoints of algebra and geometry. The relationship between chaotic homeomorphisms and chaotic foldings on chaotic connected edge graphs, as well as their fundamental group, is deduced. The fundamental group of the limit chaotic chain of foldings on chaotic is being mentioned. Different types of chaotic foldings are achieved. The theorems that govern these relations have been established. In addition, we investigate numerous applications in the domains of chemistry and biology.

REFERENCES

1. Bailey S. Dot Product Graphs and Their Applications to Ecology. Logan, Utah: Utah State University; 2013
2. Pirzada S, Dharwadker A. Applications of graph theory. The Journal of the Korean Society for Industrial and Applied Mathematics. 2007;11(4):19-38
3. Bondy JA, Murty USR. Graph Theory. Springer; 2008
4. Abu-Saleem M. Folding on the wedge sum of graphs and their fundamental group. Applied Sciences. 2010;12:14-19
5. Abu-Saleem M. Retractions and homomorphisms on some operations of graphs. Journal of Mathematics. 2018:1-4
6. Biggs NL, Lloyd EK, Wilson RJ. Graph Theory. Oxford: Oxford University Press; 1976. pp. 1736-1936
7. Chartrand G, Zhang P. Chromatic Graph Theory. First Edition. Taylor & Francis Group; 2009
8. Giblin J. Graphs, Surfaces, and Homology: An Introduction to Algebraic Topology. Second Edition. London: Chapman and Hall Ltd.; 1977
9. Rahman MS. Basic Graph Theory. Springer International Publishing AG; 2017
10. White AT. Graphs, Groups and Surfaces. Amsterdam: North-Holland Publishing Company; 1973
11. Wilson RJ, Watkins JJ. Graphs, An Introductory Approach, A First Course in Discrete Mathematics. Canada John Wiley & sons, Inc.; 1990
12. Hammack R, Imrich W, Klavžar S. Handbook of Product Graphs. Boca Raton: Taylor & Francis Group; 2011
13. Hell P, Nesetfil J. Graphs and Homomorphisms. New York: Oxford University Press; 2004
14. El-kholy E, El-Esawy A. Graph folding of some special graphs. Journal of Mathematics and Statistics. 2005;1(1):66-70
15. Massey WS. Algebraic Topology, an Introduction. New York: Harcourt Brace and World; 1967
16. Jones D. Chaos Math Online. Available from: http://www.mcasco.com [Accessed: 11 November 2001]
17. Abu-Saleem M. Folding on the chaotic Cartesian product of manifolds and their fundamental group. Tamkang Journal of Mathematics. 2008;39:353-361

18. Abu-Saleem M. Dynamical manifold and their fundamental group. Advanced Studies in Contemporary Mathematics. 2010;20(1):125-131

19. Abu-Saleem M. Conditional fractional folding of a manifold and their fundamental group. Advanced Studies in Contemporary Mathematics. 2010;20(2):271-277

20. Abu-Saleem M. Folding on manifolds and their fundamental group. Journal of Taibah University for Science. United Kingdom; 2018:1-4

21. El-Ghoul M, El-Ahmady AE, Abu-Saleem M. Folding on the Cartesian product of manifolds and their fundamental group. Applied Sciences. Romania; 2007;9:86-91

22. El-Ghoul M, El-Ahmady AE, Rafat H, Abu-Saleem M. The fundamental group of the connected sum of manifolds and their foldings. International Journal of Pure and Applied Mathematics. 005;22(2):557-567

23. Frigerio R. Hyperbolic manifold with geodesic boundary which are determine by their fundamental group. Topology and its Applications. 2004;45:69-81

24. Neto O, Silva PC. The fundamental group of an algebraic link. Comptes Rendus de l'Académie des Sciences, Paris, Série I. 2005;340:141-146

25. Abu-Saleem M. On dynamical chaotic de sitter spaces and their deformation retracts. Proceedings of the Jangjeon Mathematical Society. 2011;14:231-238

26. Abu-Saleem M. Dynamical chaotic homotopy group and its applications. International Journal of Nonlinear Science. 2011;11(12):206-212

27. Alligood KT, Sauer TD, Yorke JA. Chaos: An Introduction to Dynamical Systems. New York, Heidelberg, Berlin: Springer-Verlag; 1990

28. Devaney RL. An Introduction to Chaotic Dynamical Systems. 2nd ed. New York: Addison-Wesley; 1989

29. Peitgen H-O, Jürgen H, Saupe D. Chaos and Fractals: New Frontiers of Science. New York: Springer-Verlag; 1992

30. Rafat H. Conditional foldings of chaotic manifolds and their retractions [thesis]. Tanta, Egypt: University of Tanta; 2004

31. Kutschenreiter-Praszkiewicz I. Graph-based decision making in industry. In: Sirmacek B, editor. Graph Theory—Advanced Algorithms and Applications. IntechOpen; 2017. DOI: 10.5772/intechopen.72145. Available from: https://www.intechopen.com/books/graph-theory-advanced-algorithms-and-applications/graph-based-decision-making-in-industry

32. Mayer DC. Modeling rooted in-trees by finite p-groups. In: Sirmacek B, editor. Graph Theory—Advanced Algorithms and Applications. United Kingdom: IntechOpen; 2017. DOI: 10.5772/intechopen.68703. Available from: https://www.intechopen.com/books/graph-theory-advanced-algorithms-and-applications/modeling-rooted-in-trees-by-finite-p-groups

05 AN INVESTIGATION INTO HILBERT SPACES AND REPRODUCING KERNELS

1. INTRODUCTION: AN OVERVIEW

This chapter provides an introduction to the theory of Hilbert spaces and reproducing kernels. First, we will introduce fundamental concepts, statements, and mathematical principles from functional analysis that are specifically relevant to Hilbert spaces. The first portion covers the concepts of linear space, norm, inner product, and pre-Hilbert spaces. The second section focuses on the essential characteristics of the replicating kernels and the associated Hilbert spaces. The latter section examines the processes involving reproducing kernels, the inclusion property, the Bergman kernel, and additional qualities, along with examples of the reproducing kernels.

2. INTRODUCTION TO HILBERT SPACES

We commence by establishing the precise definition of a vector space and exploring its associated concepts. Let C denote the field of complex numbers. The subsequent introductory topics might be regarded as fundamental principles of Hilbert spaces.

2.1 Vector spaces and inner product spaces

Vector space. A vector space is a linear space that satisfies closure properties under vector addition and scalar multiplication. To be more specific, if we represent our linear space as *H* over the field C, it can be deduced that

(i) if x,y, z$\in\mathcal{H}$, then

$$x+y=y+x\in\mathcal{H},\quad x+(y+z)=(x+y)+z\in\mathcal{H};$$

(ii) if *k* is scalar, then *k* x$\in\mathcal{H}$.

Inner product. Let $\mathcal{H}$ be a linear space over the complex field £. An *inner product* on $\mathcal{H}$ is a two variable function

$$\langle\cdot,\cdot\rangle\colon \mathcal{H}\times\mathcal{H}\to C, \text{ satisfying}\cdot$$

(i) $\langle f,g\rangle = \overline{\langle g,f\rangle}$ for $f,g \in \mathcal{H}$.

(ii) $\langle \alpha f + \beta g, h\rangle = \alpha\langle f,h\rangle + \beta\langle g,h\rangle$ and $\langle f, \alpha g + \beta h\rangle = \overline{\alpha}\langle f,g\rangle + \overline{\beta}\langle f,h\rangle$ for $\alpha,\beta \in \mathbb{C}$ and $f,g,h \in \mathcal{H}$.

(iii) $\langle f,f\rangle \geq 0$ for $f \in \mathcal{H}$ and $\langle f,f\rangle = 0 \Leftrightarrow f = 0$.

Pre-Hilbert space. A *pre-Hilbert space* $\mathcal{H}$ is a linear space over the complex field $\mathbb{C}$ with an inner product defined on it.

Norm space or inner product space. A norm on an inner product space $\mathcal{H}$ denoted by $\|\cdot\|$ is defined by

$$\|f\| = \langle f,f\rangle^{1/2} \text{ or } \|f\|_{\mathcal{H}} = \langle f,f\rangle_{\mathcal{H}}^{1/2}$$

where $f\in\mathcal{H}$ and $\langle\cdot,\cdot\rangle=\langle\cdot,\cdot\rangle_{\mathcal{H}}$ denote the inner product on $\mathcal{H}$. The corresponding space is called as the inner product space or the norm space.

Properties of norm. For all $f,g \in \mathcal{H}$, and $\lambda\in\mathbb{C}$, we have

- $\|f\|\geq 0$. (Observe that the equality occurs only if $f=0$.
- $\|\lambda f\|=|\lambda|\,\|f\|$.

Schwarz inequality. For all $f,g\in\mathcal{H}$, it follows that

$$|\langle f,g\rangle| \leq \|f\|\|g\|. \tag{1}$$

In case if f and g are linearly dependent, then the inequality becomes equality.

Triangle inequality. For all $f,g \in \mathcal{H}$, it follows that

$$\|f+g\| \leq \|f\| + \|g\|. \tag{2}$$

In case if f and g are linearly dependent, then the inequality becomes equality.

Polarization identity. For all $f,g \in \mathcal{H}$, it follows that

$$\langle f,g\rangle = \frac{1}{4}(\|f+g\|^2 - \|f-g\|^2 + i\|f+ig\|^2 - \|f-ig\|^2) \text{ for } f,g \in \mathcal{H}. \tag{3}$$

Parallelogram identity. For all $f,g \in\mathcal{H}$, it follows that

$$\|f+g\|^2 + \|f-g\|^2 = 2\|f\|^2 + 2\|g\|^2. \tag{4}$$

Metric. A metric on a set X is a function $d:X\times X\to\mathbb{R}$ satisfying the properties.

- $d(x,y)\geq 0$ and $d(x,y)=0$ only if $x=y$;
- $d(x,y)=d(y,x)$;
- $d(x,y)\leq d(x,z)+d(z,y)$;

for all $x,y,z\in X$. Moreover the space (X,d) is the associated metric space. If we rearrange the metric with its properties for the inner product space $\mathcal{H}$, then it follows that for all $f,g, h \in \mathcal{H}$ and for all $\lambda \in \mathbb{C}$, where d satisfies all requirements to be a metric, we have

- $d(f,g) \geq 0$ and equality occurs only if $f=g$.
- $d(f,g)=d(g,f)$.
- $d(f,g) \leq d(f,h)+d(h,g)$.
- $d(f-h,g-h)=d(f,g)$.
- $d(\lambda f,\lambda g)=|\lambda| \cdot d(f,g)$.

Note. The binary function d given in the metric definition above represents the metric topology in $\mathcal{H}$ which is called *strong topology* or *norm topology.* As a result, a sequence $(f)_{n \geq 0}$ in the pre-Hilbert space $\mathcal{H}$ converges strongly to f if the condition

$$\|f_n - f\| \to 0 \text{ whenever } n \to \infty$$

is satisfied.

2.2 Introduction to linear operators

Linear operator. A map L from a linear space to another linear space is called *linear operator* if

$$L(\alpha f + \beta g) = \alpha Lf + \beta Lg$$

is satisfied for all $\alpha, \beta \in \mathbb{C}$, and for all $f,g \in \mathcal{H}$.

Continuous operator. An operator L is said to be continuous if it is continuous at each point of its domain. Notice that the domain and range spaces must be convenient for appropriate topologies.

Lipschitz constant of a linear operator. If L is a linear operator from $\mathcal{H}$ to $\mathcal{G}$ where $\mathcal{H}$ and $\mathcal{G}$ are pre-Hilbert spaces, then the Lipschitz constant for L is its norm $\|L\|$ and it is defined by

$$\|L\| = \sup\{\|Lf\|_{\mathcal{G}} / \|f\|_{\mathcal{H}} : 0 \neq f \in \mathcal{H}\}. \quad (5)$$

Theorem 1. Let L be a linear operator from the pre-Hilbert spaces $\mathcal{H}$ to $\mathcal{G}$. Then the followings are mutually equivalent:

(i) L is continuous.

(ii) is bounded, that is,

$$\sup\{\|Lf\|_G : \|f\|_H \leq k\} < \infty$$

for $0 \leq k < \infty$.

(iii) L is Lipschitz continuous, that is,

$$\|Lf - Lg\|_G \leq \lambda \|f - g\|_{\mathcal{H}},$$

where $0 \leq \lambda < \infty$ and $f,g \in H$.

Some properties of linear operators. Let B($\mathcal{H}$,$\mathcal{G}$) be the collection of all continuous linear operators from the pre-Hilbert spaces $\mathcal{H}$ to $\mathcal{G}$. Then

- ⦿ B($\mathcal{H}$,$\mathcal{G}$) is a linear space with respect to the natural addition and scalar multiplication satisfying

$$(\alpha L + \beta M)f = \alpha Lf + \beta Mf,$$

where L and M are linear operators, $f \in \mathcal{H}$ and $\alpha, \beta \in \mathbb{C}$.

- ⦿ Whenever $\mathcal{H}=\mathcal{G}$, then B($\mathcal{H}$,$\mathcal{G}$) Is denoted by B(H).
- ⦿ If K is another pre-Hilbert space, L∈B($\mathcal{H}$,$\mathcal{G}$) and K∈B($\mathcal{G}$,$\mathcal{K}$). Then the product

$$(KL)f = K(Lf) \text{ for } f \in \mathcal{H} \in B(\mathcal{H}, \mathcal{K}).$$

In addition,

(i) $K(\xi L+\zeta M)=\xi KL+\zeta KM$

(ii) $\|\xi L\|=|\xi|\cdot\|L\|$

(iii) $\|L+M\|\leq\|L\|+\|M\|$ and

(iv) $\|KL\|\leq\|K\|\|L\|$.

are also satisfied.

2.3 Hilbert spaces and linear operators

Linear form (or linear functional). A linear operator from the pre-Hilbert space $\mathcal{H}$ to the scalar field $\mathbb{C}$ is called a *linear form* (or *linear functional*).

Hilbert spaces. A pre-Hilbert space $\mathcal{H}$ is said to be a *Hilbert space* if it is complete in metric. In other words if f_n is a Cauchy sequence in $\mathcal{H}$, that is, if

$$\|f_n - f_m\| \to 0 \text{ whenever } n, m \to \infty,$$

then there is $f \in \mathcal{H}$ such that

$$\|f_n - f\| \to 0 \text{ whenever } n \to \infty.$$

Note. Every subspace of a pre-Hilbert space is also a pre-Hilbert space with respect to the induced inner product. However, the reverse is not always true. For a subspace of a Hilbert space to be also a Hilbert space, it must be closed.

Completion. The canonical method for which a pre-Hilbert space $\mathcal{H}$ is embedded as a dense subspace of a Hilbert space $\tilde{\mathcal{H}}$ so that

$$\langle f, g\rangle_{\tilde{\mathcal{H}}} = \langle f, g\rangle_{\mathcal{H}} \text{ for } f, g \in \mathcal{H}$$

is called *completion*.

Note. If L is a continuous linear operator from a dense subspace $\mathcal{M}$ of a Hilbert space $\mathcal{H}$ to a Hilbert space $\mathcal{G}$, then it can be extended uniquely to a continuous linear operator from $\mathcal{H}$ to $\mathcal{G}$ with preserving norm.

Theorem 2. Let $\mathcal{M}$ and $\mathcal{N}$ be dense subspaces of the Hilbert spaces $\mathcal{H}$ and $\mathcal{G}$, respectively. For $f \in \mathcal{H}$, $g \in \mathcal{M}$ and $0 \leq \lambda < \infty$, if a linear operator L from $\mathcal{M}$ to $\mathcal{G}$ satisfies

$$|\langle Lf, g\rangle_{\mathcal{G}}| \leq |\lambda| \|f\|_{\mathcal{H}} \|g\|_{\mathcal{G}}, \tag{6}$$

then L is uniquely extended to a continuous linear operator from $\mathcal{M}$ to $\mathcal{G}$ with norm $\leq \lambda$ where the norm coincides with the minimum of such λ.

Theorem 3. Let (Ω, μ) denotes a measure space so that Ω is the union of subsets of finite positive measure and $L^2(\Omega, \mu)$ consists of all measurable functions $f(\omega)$ on Ω such that

$$\int_{\Omega} |f(\omega)|^2 d\mu(\omega) < \infty. \tag{7}$$

Then $L^2(\Omega, \mu)$ is a Hilbert space with respect to the inner product

$$\langle f, g\rangle := \int_{\Omega} f(\omega)\overline{g(\omega)}\, d\mu(\omega). \tag{8}$$

Theorem 4 (F. Riesz). For each continuous linear functional φ on a Hilbert space $\mathcal{H}$, there exists uniquely $g \in \mathcal{H}$ such that

$$\varphi(f) = \langle f, g\rangle \text{ for } f \in \mathcal{H}. \tag{9}$$

Theorem 5. Let $\mathcal{M}$ be a closed subspace of a Hilbert space $\mathcal{H}$. Then the algebraic direct sum relation

$$\mathcal{H} = \mathcal{M} \oplus \mathcal{M}^{\perp}$$

is satisfied. In other words $\forall f \in \mathcal{H}$ can be uniquely written by

$$f = f_{\mathcal{M}} + f_{\mathcal{M}^{\perp}} \text{ with } f_{\mathcal{M}} \in \mathcal{M},\ f_{\mathcal{M}^{\perp}} \in \mathcal{M}^{\perp}. \tag{10}$$

In addition, $\|f_{\mathcal{M}}\|$ coincides with the distance from f to $\mathcal{M}^{\perp}$

$$\|f_{\mathcal{M}}\| = \min\{\|f - g\| : g \in \mathcal{M}^{\perp}\}. \tag{11}$$

Remark. In a Hilbert space, the closed linear span of any subset $\mathcal{A}$ of a Hilbert space $\mathcal{H}$ coincides with $(A^{\perp})^{\perp}$.

Total subset of a Hilbert space. A subset $\mathcal{A}$ of a Hilbert space $\mathcal{H}$ is called *total* in $\mathcal{H}$ if 00 is the only element that is orthogonal to all elements of $\mathcal{A}$. In other words,

$$\mathcal{A}^{\perp} = \{0\}.$$

As a result, $\mathcal{A}$ is total if and only if every element of $\mathcal{H}$ can be approximated by linear combinations of elements of $\mathcal{A}$.

Orthogonal projection. If $\mathcal{M}$ is a closed subspace of $\mathcal{H}$, the map $f \mapsto f_{\mathcal{M}}$ gives a linear operator from $\mathcal{H}$ to $\mathcal{M}$ with norm ≤1. We call this operator as the *orthogonal projection* to $\mathcal{M}$ and denote it by $P_{\mathcal{M}}$.

Note. If I is the identity operator on $\mathcal{H}$, then $I\text{-}P_{\mathcal{M}}$ denotes the orthogonal projection to $\mathcal{M}^{\perp}$, and the relation

$$\|f\|^2 = \|P_{\mathcal{M}}f\|^2 + \|(I - P_{\mathcal{M}})f\|^2 \tag{12}$$

is satisfied for all $f \in \mathcal{H}$.

Weak topology. The weakest topology that makes continuous all linear functionals of the form $f \mapsto \langle f,g \rangle$ is called *the weak topology* of a Hilbert space $\mathcal{H}$.

Note. If $f \in \mathcal{H}$, then with respect to the weak topology, a fundamental system of neighborhoods of f is composed of subsets of the form

$$U(f;A,\epsilon) = \{h : |\langle f,g \rangle - \langle h,g \rangle| < \epsilon \text{ for } g \in \mathcal{A}\},$$

where $\mathcal{A}$ is a finite subset of $\mathcal{H}$ and $\epsilon > 0$. Then a directed net $\{f_\lambda\}$ converges weakly to f if and only if

$$\langle f_\lambda, g \rangle \xrightarrow{\lambda} \langle f,g \rangle \text{ for all } g \in \mathcal{H}.$$

Operator weak topology. The weakest topology that makes continuous all linear functionals of the form

$$L \mapsto \langle Lf,g \rangle \text{ for } f \in \mathcal{H}, g \in \mathcal{G}$$

is called *the operator weak topology* in the space $B(\mathcal{H},\mathcal{G})$ of continuous linear operators from $\mathcal{H}$ to $\mathcal{G}$. In addition, a directed net $\{L_\lambda\}$ converges weakly to L if

$$\langle L_\lambda f, g \rangle \xrightarrow{\lambda} \langle Lf, g \rangle.$$

Operator strong topology. The weakest topology that makes continuous all linear operators of the form

$$L \mapsto Lf \text{ for } f \in \mathcal{H}$$

is called *the operator strong topology*. Moreover a directed net $\{L_\lambda\}$ converges strongly to L if

$$\|L_\lambda f - Lf\| \xrightarrow{\lambda} 0 \text{ for all } f \in \mathcal{H}.$$

Theorem 6. Let $\mathcal{H}$ and $\mathcal{G}$ be Hilbert spaces and $B(\mathcal{H},\mathcal{G})$ be a continuous linear operator from $\mathcal{H}$ to $\mathcal{G}$. Then

- The closed unit ball $U := \{f : \|f\| \leq 1\}$ of $\mathcal{H}$ is weakly compact;
- The closed unit ball $\{L : \|L\| \leq 1\}$ of $B(\mathcal{H},\mathcal{G})$ is weakly compact.

Theorem 7. Let $\mathcal{H}$ be a Hilbert space and $A \subseteq \mathcal{H}$. Then if A is weakly bounded in the sense

$$\sup_{f\in A}|\langle f,g\rangle| < \infty \text{ for } g\in \mathcal{H}, \tag{13}$$

then it is strongly bounded, that is, $\sup_{f\in A}\|f\| < \infty$.

Theorem 8. If $\mathcal{H}$ and $\mathcal{G}$ are Hilbert spaces and L is a linear operator from $\mathcal{H}$ to $\mathcal{G}$, then the strong continuity and weak continuity for L are equivalent.

Theorem 9. Let $\mathcal{H}$ and $\mathcal{G}$ be Hilbert spaces. Then the following statements for $\mathrm{L}\subseteq B(\mathcal{H},\mathcal{G})$ are mutually equivalent:

1. L is weakly bounded; that is, for $f\in\mathcal{H}$, $g\in\mathcal{G}$, we have
$$\sup_{L\in \mathrm{L}}|\langle Lf,g\rangle| < \infty$$
2. L is strongly bounded; that is, for $f\in\mathcal{H}$, we have
$$\sup_{L\in \mathrm{L}}\|Lf\| < \infty.$$
3. L is norm bounded (or uniformly bounded); that is,
$$\sup_{L\in \mathrm{L}}\|L\| < \infty.$$

Theorem 10. A linear operator L from the Hilbert spaces $\mathcal{H}$ to $\mathcal{G}$ is said to be *closed* if its *graph*

$$G_L := \{f\oplus Lf : f\in\mathcal{H}\} \tag{14}$$

is a closed subspace of the direct sum space $\mathcal{H}\oplus\mathcal{G}$, that is, whenever $n\to\infty$,

$$\|f_n - f\| \to 0 \textit{ in } \mathcal{H} \textit{ and } \|Lf_n - g\| \to 0 \textit{ in } \mathcal{G} \Rightarrow g = Lf.$$

Theorem 11. If L is a closed linear operator with a domain of a Hilbert space H to another Hilbert space G, then it is continuous.

Sesqui-linear form. A function $\Phi: \mathcal{H}\times\mathcal{G}\to\mathbb{C}$ is a *sesqui-linear form* (or *sesquilinear function*) if for $f,h\in\mathcal{H}, g,k\in\mathcal{G}$ and $\alpha,\beta\in\mathbb{C}$,

$$\text{(i) } \Phi(\alpha f+\beta h,g) = \alpha\Phi(f,g)+\beta\Phi(h,g) \tag{15}$$

$$\text{(ii) } \Phi(f,\alpha g+\beta k) = \overline{\alpha}\Phi(f,g)+\overline{\beta}\Phi(f,k) \tag{16}$$

are satisfied where $\mathcal{H}$ and $\mathcal{G}$ are Hilbert spaces.

Remark. If $L\in B(\mathcal{H},\mathcal{G})$, then the sesqui-linear form Φ defined by

$$\Phi(f,g) = \langle Lf,g\rangle_G \tag{17}$$

is bounded in the sense that

$$|\Phi(f,g)| \le \lambda\|f\|_{\mathcal{H}}\|g\|_{\mathcal{G}} \text{ for } f\in\mathcal{H},\ g\in\mathcal{G}, \tag{18}$$

Where $\lambda \ge \|L\|$

Remark. If a sesqui-linear form Φ satisfies the condition (18), then for $f\in\mathcal{H}$, the linear functional

$$g\mapsto\overline{\Phi(f,g)}$$

is continuous on $\mathcal{G}$. If we apply the Riesz theorem, then there exists uniquely $f'\in\mathcal{G}$ satisfying

$$\|f'\|_{\mathcal{G}}\leq\lambda\|f\|_{\mathcal{H}} \text{ and } \Phi(f,g)=\langle f',g\rangle_{\mathcal{G}} \text{ for } g\in\mathcal{G}.$$

Hence $f\mapsto f'$ becomes linear, and as a result we obtain

$$\Phi(f,g)=\langle f',g\rangle_{\mathcal{G}}=\langle Lf,g\rangle_{\mathcal{G}}.$$

Adjoint operator. If $L\in B(\mathcal{H},\mathcal{G})$, then the unique operator $L^*\in B(\mathcal{G},\mathcal{H})$ satisfying

$$\Phi(f,g)=\langle f,L^*g\rangle_{\mathcal{H}} \text{ for } f\in\mathcal{H}, g\in\mathcal{G} \tag{19}$$

is called the *adjoint* of L.

Remark. By the definitions of L and L^*, it follows that

$$\langle Lf,g\rangle_{\mathcal{G}}=\langle f,L^*g\rangle_{\mathcal{H}} \text{ for } f\in\mathcal{H},\ g\in\mathcal{G}. \tag{20}$$

Isometric property. The adjoint operation is isometric if

$$\|L\|=\|L^*\| \text{ is satisfied.} \tag{21}$$

Remark. Let $\mathcal{H},\mathcal{G}$, and K be Hilbert spaces and $K\in B(\mathcal{G},\mathcal{K})$ and $L\in B(\mathcal{H},\mathcal{G})$ be given. Then

$$KL\in B(\mathcal{H},\mathcal{K}) \text{ and } (KL)^*=L^*K^* \tag{22}$$

$$\text{Ker}(L)=(\text{Ran}(L^*))^{\perp} \text{ and } (\text{Ker}(L))^{\perp}=\text{Clos}\{\text{Ran}(L)^*\} \tag{23}$$

where Ker(L) is the kernel of L and Ran(L) is the range of L.

Theorem 12. If $L,M\in B$ $(\mathcal{H},\mathcal{G})$, then the following statements are mutually equivalent.

(i) Ran(M) $\subseteq$Ran(L).

(ii) There exists $K\in B(\mathcal{H})$ such that $M=LK$.

(iii) There exists $0\leq\lambda<\infty$ such that

$$\|M^*g\|\leq\lambda\|L^*g\| \text{ for } g\in\mathcal{G}.$$

Quadric form. Let $\mathcal{H}$ be a Hilbert space. A function

$$\varphi:\mathcal{H}\to\mathbb{C}$$

is a *quadratic form* if for all $f\in\mathcal{H}$ and $\zeta\in\mathbb{C}$,

$$\varphi(\zeta f)=|\zeta|^2\varphi(f) \tag{24}$$

and

$$\varphi(f+g)+\varphi(f-g)=2\{\varphi(f)+\varphi(g)\} \tag{25}$$

are satisfied.

Note. If $L\in B(\mathcal{H})$, the quadratic form φ on $\mathcal{H}$ is defined by

$$\varphi(f) = \langle Lf, f\rangle \text{ for } f \in \mathcal{H}, \tag{26}$$

and it is bounded

$$|\varphi(f)| \leq \lambda \|f\|^2 \text{ for } f \in \mathcal{H}, \tag{27}$$

where $\lambda \geq \|L\|$.

Remark. The sesqui-linear form Φ associated with L can be recovered from the quadratic form φ by the equation

$$\Phi(f,g) = \frac{1}{4}\{\varphi(f+g) - \varphi(f-g)\} + \{\varphi(f+ig) - \varphi(f-ig)\} \tag{28}$$

for all $f,g \in \mathcal{H}$.

Self-adjoint operator. A continuous linear operator L on a Hilbert space H is said to be *self-adjoint* if $L=L^*$.

Remark. L is self-adjoint if and only if the associated sesqui-linear form Φ is Hermitian.

Remark. If L is self-adjoint, then the norm of L coincides with the minimum of λ given in (27) for the related quadratic form

Theorem 13. If L is a continuous self-adjoint operator, then

$$\|L\| = \sup\{|\langle Lf, f\rangle| : \|f\| \leq 1\}. \tag{29}$$

Positive definite operator. A self-adjoint operator $L\in B(\mathcal{H})$ is said to be *positive* (or *positive definite*) if

$$\langle Lf, f\rangle \geq 0 \quad \text{for all } f \in \mathcal{H}.$$

If $\langle Lf,f\rangle=0$ only when $f=0$, then L is said to be *strictly positive* (or, *strictly positive definite*).

Note. For any positive operator $L\in B(\mathcal{H})$, the Schwarz inequality holds in the following sense

$$|\langle Lf, g\rangle|^2 \leq \langle Lf, f\rangle \cdot \langle Lg, g\rangle. \tag{30}$$

Theorem 14. Let L and M be continuous positive operators on $\mathcal{H}$ and $\mathcal{G}$, respectively. Then a continuous linear operator K from $\mathcal{H}$ to $\mathcal{G}$ satisfies the inequality

$$|\langle Kf, g\rangle_{\mathcal{G}}|^2 \leq \langle Lf, f\rangle_{\mathcal{H}} \langle Mg, g\rangle_{\mathcal{G}} \text{ for } f \in \mathcal{H}, \in \mathcal{G} \tag{31}$$

if and only if the continuous linear operator

$$\begin{bmatrix} L & K^* \\ K & M \end{bmatrix}$$

on the direct sum Hilbert space $\mathcal{H}\oplus\mathcal{G}$ with

$$f \oplus g \mapsto (Lf + K^* g) \oplus (Kf + Mg)$$

is positive definite.

Theorem 15. Let L be a continuous positive definite operator. Then there exists a unique positive definite operator called the *square root* of L, denoted by $L^{1/2}$, such that $(L^{1/2})^2=L$.

Modulus operator. The square root of the positive definite operator L^*L is called the *modulus* (*operator*) of L if L is a continuous linear operator.

Isometry. A linear operator U between Hilbert spaces $\mathcal{H}$ and $\mathcal{G}$ is called *isometric* or an *isometry* if

$$\|Uf\|_{\mathcal{G}} = \|f\|_{\mathcal{H}} \text{ for } f \in \mathcal{H} \tag{32}$$

is satisfied, that is, it preserves the norm.

Note. Eq. (32) implies that a continuous linear operator U is isometric if and only if $U^*U=I_{\mathcal{H}}$; in other words,

$$\langle Uf, Ug\rangle_{\mathcal{G}} = \langle f,g\rangle_{\mathcal{H}} \text{ for } f,g \in \mathcal{H}, \tag{33}$$

that is, U preserves the inner product.

Unitary operator. A surjective isometry linear operator U: $\mathcal{H}\rightarrow \mathcal{H}$ is called a *unitary* (*operator*).

Note. Observe that if $U \in B(\mathcal{H})$ is a unitary operator, then $U^*=U^{-1}$.

Partial isometry. A continuous linear operator U between Hilbert spaces $\mathcal{H}$ and $\mathcal{G}$ is called a *partial isometry* if

$$f \in (\mathrm{Ker}U)^{\perp} = \mathrm{Ran}(U^*) \Rightarrow \|Uf\| = \|f\|.$$

The spaces $(\mathrm{Ker}U)^{\perp}$ and Ran(U) are called the *initial space* of U and the *final space* of U, respectively.

Note. If U is a partial isometry, then its adjoint U is also a partial isometry.

Theorem 17. Every continuous linear operator L on $\mathcal{H}$ admits a unique decomposition

$$L = U\tilde{L}, \tag{34}$$

where $\tilde{L}$ is a positive definite operator and U is a partial isometry with initial space the closure of Ran($\tilde{L}$).

3. REPRODUCING KERNELS AND RKHS

We continue our analysis on the abstract theory of reproducing kernels.

3.1 Definition and fundamental properties

Reproducing kernels. Let $\mathcal{H}$ be a Hilbert space of functions on a nonempty set X with the inner product $\langle f,g\rangle$ and norm $\|f\|=\langle f,f\rangle^{1/2}$ for f and $g\in\mathcal{H}$ and. Then the complex valued function $K(y,x)$ of y and x in X is called a *reproducing kernel of* $\mathcal{H}$ if

1. For all $x\in X$, it follows that $K_x(\cdot)=K(\cdot,x)\in\mathcal{H}$,
2. For all $x\in X$ and all $f\in\mathcal{H}$,

$$f(x) = \langle f, K_x\rangle, \tag{35}$$

are satisfied.

Note. Let K be a reproducing kernel. Applying (35) to the function K_x at y, we get

$$K_x(y) = K(y,x) = \langle K_y, K_x\rangle, \text{ for } x,y\in X. \tag{36}$$

Then, for any $x\in X$, we obtain

$$\|K_x\| = \langle K_x, K_x\rangle^{1/2} = K(x,x)^{1/2}. \tag{37}$$

Note. Observe that the subset $\{K_x\}_{x\in X}$ is total in $\mathcal{H}$, that is, its closed linear span coincides with $\mathcal{H}$. This follows from the fact that, if $f\in\mathcal{H}$ and $f\perp K_x$ for all $x\in X$, then

$$f(x) = \langle f, K_x\rangle = 0 \text{ for all } x\in X,$$

and hence f is the 00 element in $\mathcal{H}$. As a result, $\{0\}^{\perp}=\mathcal{H}$.

RKHS. A Hilbert space $\mathcal{H}$ of functions on a set X is called a RKHS if there exists a reproducing kernel K of $\mathcal{H}$.

Theorem 18. If a Hilbert space $\mathcal{H}$ of functions on a set X admits a reproducing kernel K, then this reproducing kernel K is unique.

Theorem 19. There exists a reproducing kernel K for $\mathcal{H}$ for a Hilbert space $\mathcal{H}$ of functions on X, if and only if for all $x\in X$, the linear functional $H\ni f\mapsto f(x)$ of evaluation at x is bounded on $\mathcal{H}$.

Hermitian and positive definite kernel. Let X be an arbitrary set and K be a kernel on X, that is, $K:X\times X\to\mathbb{C}$. The kernel K is called *Hermitian* if for any finite set of points $\{y_1,\ldots,y_n\}\subseteq X$, we have

$$\sum_{i,j=1}^{n} \overline{\epsilon_j}\epsilon_i K(y_j,y_i) \in \mathbb{R}.$$

It is called *positive definite*, if for any complex numbers $\epsilon_1,\ldots,\epsilon_n$, we have

$$\sum_{i,j=1}^{n} \overline{\epsilon_j}\epsilon_i K(y_j,y_i) \geq 0.$$

Note. From the previous inequality, it follows that for any finitely supported family of complex numbers $\{\epsilon_x\}_{x\in X}$, we have

$$\sum_{x,y\in X} \overline{\epsilon}_y \epsilon_x K(y,x) \geq 0.$$

Theorem 20. The reproducing kernel k of a reproducing kernel Hilbert space $\mathcal{H}$ is a positive definite matrix in the sense of E.H. Moore.

Properties of RKHS. Given a reproducing kernel Hilbert space $\mathcal{H}$ and its kernel $K(y,x)$ on X, then for all $x,y \in X$, we have

(i) $K(y,y)\geq 0$.

(ii) $K(y,x)= \overline{K(x,y)}$.

(iii) $|K(y,x)|^2 \leq K(y,y)K(x,x)$ (Schwarz inequality).

(iv) Let $x_0 \in X$. Then the following statements are equivalent:

(a) $K(x_0,x_0)=0$.

(b) $K(y,x_0)=0$ for all $y \in X$.

(c) $f(x_0)=0$ for all $f \in \mathcal{H}$.

Theorem 21. For any positive definite kernel K on X, there exists a unique Hilbert space $\mathcal{H}_K$ of functions on X with reproducing kernel K.

Theorem 22. Every sequence of functions $(f_n)_{n\geq 1}$ that converges strongly to a function f in $\mathcal{H}_K(X)$ converges also in the pointwise sense, i.e., for any point $x \in X$,

$$\lim_{n\to\infty} f_n(x) = f(x).$$

In addition, this convergence is uniform on every subset of X on which $x \mapsto K(x,x)$ is bounded.

Theorem 23. A complex valued function g on X belongs to the reproducing kernel Hilbert space $\mathcal{H}_K(X)$ if and only if there exists $0\leq\lambda<\infty$ such that,

$$\left[g(y)\overline{g(x)}\right] \leq \lambda^2 [K(y,x)] \text{ on } X. \tag{39}$$

$\|g\|$ coincides with the minimum of all such λ.

Theorem 24. If $K^{(1)}(y,x)$ and $K^{(2)}(y,x)$ are two positive definite kernels on X, then the following statements are mutually equivalent:

1. $\mathcal{H}_{K^{(1)}}(X) \subseteq \mathcal{H}_{K^{(2)}}(X)$.

2. There exists $0\leq\lambda<\infty$ such that

$$\left[K^{(1)}(y,x)\right] \leq \lambda^2 \left[K^{(2)}(y,x)\right].$$

Note. For any map φ from a set X to a Hilbert space $\mathcal{H}$, with the notation $x \mapsto \varphi_x$, a kernel K can be defined by

$$K(y,x) = \left\langle \varphi_x, \varphi_y \right\rangle \text{ for } x, y \in X. \tag{40}$$

Theorem 25. Let $\varphi: X \mapsto \mathcal{H}$ be an arbitrary map and for $x,y \in X$ let K be defined as

$$K(y,x) = \left\langle \varphi_x, \varphi_y \right\rangle.$$

Then K is a positive definite kernel.

Theorem 26. Let T be the linear operator from $\mathcal{H}$ to the space of functions on X, defined by

$$(Tf)(x) = \langle f, \varphi_x \rangle \text{ for } x \in X, f \in \mathcal{H}.$$

Then Ran(T) coincides with $\mathcal{H}_K(X)$ and

$$\|Tf\|_K = \|P_{\mathcal{M}} f\| \text{ for } f \in \mathcal{H},$$

where $\mathcal{M}$ is the orthogonal complement of Ker(T), $P_{\mathcal{M}}$ is the orthogonal projection onto $\mathcal{M}$, and $\|\cdot\|_K$ denotes the norm in $\mathcal{H}_K(X)$.

Kolmogorov decomposition. Let $K(y,x)$ be a positive definite kernel on an abstract set X. Then there exists a Hilbert space $\mathcal{H}$ and a function $\varphi: X \to \mathcal{H}$ such that

$$K(y,x) = \left\langle \varphi_x, \varphi_y \right\rangle \text{ for } x, y \in X.$$

3.2 Operations with RKHSs

Theorem 27. Let $K^{(0)}$ be the restriction of the positive definite kernel K to a nonempty subset X_0 of X and let $\mathcal{H}_{K^{(0)}}$ and $\mathcal{H}_K(X)$ be the RKHS corresponding to $K^{(0)}$ and K, respectively. Then

$$\mathcal{H}_{K^{(0)}}(X_0) = \{ f|_{X_0} : f \in \mathcal{H}_K(X) \} \tag{41}$$

and

$$\|h\|_{K^{(0)}} = \min\left\{ \|f\|_K : f|_{X_0} = h \right\} \text{ for all } h \in \mathcal{H}_{K^{(0)}}(X_0). \tag{42}$$

Remark. If $K^{(1)}(y,x)$ and $K^{(2)}(y,x)$ are two positive definite kernels, then

$$K(y,x) = K^{(1)}(y,x) + K^{(2)}(y,x)$$

is also a positive definite kernel.

Remark. Let $\mathcal{H}_{K^{(1)}}, \mathcal{H}_{K^{(2)}}$, and $\mathcal{H}_K$ be RKHSs with reproducing kernels $K^{(1)}(y,x)$, $K^{(2)}(y,x)$, and $K(y,x)$, respectively, and let $K = K^{(1)} + K^{(2)}$. Then

$$\mathcal{H}_K(X) = \mathcal{H}_{K^{(1)}}(X) + \mathcal{H}_{K^{(2)}}(X),$$

and for $f \in \mathcal{H}_{K^{(1)}}(X)$ and $g \in \mathcal{H}_{K^{(2)}}(X)$, it follows that

Theorem 28. The intersection $\mathcal{H}_{K(1)}(X)\cap\mathcal{H}_{K(2)}(X)$ of Hilbert spaces $\mathcal{H}_{K(1)}(X)$ and $\mathcal{H}_{K(2)}(X)$ is again a Hilbert space of functions on X with respect to the norm

$$\|f\|^2 := \|f\|^2_{K^{(1)}} + \|f\|^2_{K^{(2)}}.$$

In addition the intersection Hilbert space is a RKHS.

Theorem 29. The reproducing kernel of the space

$$\mathcal{H}_K(X) = \mathcal{H}_{K^{(1)}}(X)\cap\mathcal{H}_{K^{(2)}}(X)$$

is determined, as a quadratic form, by

$$\sum_{x,y}\overline{\varepsilon_y}\varepsilon_x K(y,x) = \inf\{\sum_{x,y}\overline{\eta_y}\eta_x K^{(1)}(y,x) + \sum_{x,y}\overline{\zeta_y}\zeta_x K^{(2)}(y,x) : [\varepsilon_x] = [\eta_x] + [\zeta_x]\},$$

where $[\epsilon_x], [\eta_x], [\zeta_x]$ are an arbitrary complex valued function on X with finite support.

Theorem 30. The tensor product Hilbert space

$$\mathcal{H}_{K^{(1)}}(X) \otimes \mathcal{H}_{K^{(2)}}(X)$$

is a RKHS on X×X.

Theorem 31. The RKHS $\mathcal{H}_K(X)$ of the kernel $K(y,x)=K^{(1)}(y,x)\cdot K^{(2)}(y,x)$ consists of all functions f on X for which there are sequences $(g_n)_{n\geq 0}$ of functions in $\mathcal{H}_{K(1)}(X)$ and $(h_n)_{n\geq 0}$ of functions in $\mathcal{H}_{K(2)}(X)$ so that

$$\sum_1^\infty \|g_n\|^2_{K^{(1)}}\|h_n\|^2_{K^{(2)}} < \infty, \quad \sum_1^\infty g_n(x)h_n(x) = f(x),\ x\in X, \tag{44}$$

and the norm is given by

$$\|f\|^2_K = \min\left\{\sum_1^\infty \|g_n\|^2_{K^{(1)}}\|h_n\|^2_{K^{(2)}}\right\},$$

where the minimum is taken over the set of all sequences $(g_n)_{n\geq 0}$ and $(h_n)_{\geq 0}$ satisfying (44).

3.3 Examples of RKHS. Bergman and Hardy spaces

Bergman space. The space of all analytic functions f on Ω for which

$$\iint_\Omega |f(z)|^2 dxdy < \infty, \qquad (z = x + iy)$$

is satisfied is called the *Bergman space* on Ω and denoted by $A^2(\Omega)$.

Remark. $A^2(\Omega)$ is a *RKHS* with respect to the inner product

$$\langle f,g\rangle \equiv \langle f,g\rangle_\Omega := \iint_\Omega f(z)\overline{g(z)}dxdy,$$

and its kernel is called the *Bergman kernel* on Ω and denoted by $B^{(\Omega)}(w,z)$.

Bergman kernel for the unit disc. The Bergman kernel for the open unit disc $\mathbb{D}$ is given by

$$B^{(\mathbb{D})}(w,z) = \frac{1}{\pi}\frac{1}{(1-w\bar{z})^2} \qquad \text{for } w,z \in \mathbb{D}. \tag{45}$$

Bergman kernel of a simply connected domain. The Bergman kernel of a simply connected domain Ω (≠ℂ) is given by

$$B^{(\Omega)}(w,z) = \frac{1}{\pi}\frac{\varphi'(w)\overline{\varphi'(z)}}{\left(1-\varphi(w)\overline{\varphi(z)}\right)^2} \text{ for } w,z \in \Omega, \tag{46}$$

where φ is any conformal mapping function from Ω onto $\mathbb{D}\mathbb{D}$.

Theorem 32. A conformal mapping from Ω to $\mathbb{D}\mathbb{D}$ can be recovered from the Bergman kernel of Ω.

Jordan curve. A *Jordan curve* is a continuous 1-1 image of $\{|\xi|=1\}$ in ℂ.

Green function. A *Green function* $G(w,z)$ of Ω is a function harmonic in Ω except at z, where it has logarithmic singularity, and continuous in the closure $\bar{\Omega}$, with boundary values $G(w,z)=0$ for all $w\in\partial\Omega$, where Ω is a finitely connected domain of the complex plane.

Theorem 33. Let Ω be a finitely connected domain bounded by analytic Jordan curves, and let $G(w,z)$ be the Green's function of Ω. Then the Bergman kernel function is

$$B^{(\Omega)}(w,z) = -\frac{2}{\pi}\frac{\partial^2 G}{\partial w \partial \bar{z}}(w,z),\ w \neq z. \tag{47}$$

Hardy space. The closed linear span of $\{\varphi_n: n=0,1,...\}$ in $L^2((T)$ is called the (Hilbert type) *Hardy space* on $\mathbb{T}\mathbb{T}$ and is denoted by $H^2(\mathbb{T}\mathbb{T})$. Here $\varphi_n(\xi)=\xi^n$.

Remark. $f\in L^2(\mathbb{T}\mathbb{T})$ belongs to the Hardy space $H^2(\mathbb{T}\mathbb{T})$ if and only if it is orthonormal to all φ_n $(n<0)$, that is, all Fourier coefficients of f with negative indices vanish. Then we have

$$\langle f,g\rangle_{L^2} = \sum_{n=0}^{\infty} a_n\bar{b}_n \text{ for } f,g \in H^2(\mathbb{T}), \tag{48}$$

where

$$a_n = \langle f,\varphi_n\rangle_{L^2} \text{ and } b_n = \langle g,\varphi_n\rangle_{L^2} \quad (n = 0, 1, \ldots).$$

Szegö kernel. The kernel $S(\xi,z) := \frac{1}{1-\xi\bar{z}}$ for $\xi\in\mathbb{T}, z\in\mathbb{D}$, or its analytic extension $\tilde{S}(w,z) := \frac{1}{1-w\bar{z}}$ for $w,z \in \mathbb{D}$ is called the *Szegö kernel.*

4. AT A GLANCE

This chapter aims to offer a concise overview of Hilbert space, highlighting its essential characteristics, and to introduce the reproducing kernels of the associated spaces. We divide our analysis into two distinct sections. The initial section provides an overview of fundamental concepts in inner product spaces, encompassing the concept of norms, pre-Hilbert spaces, and ultimately Hilbert spaces. The second section focuses on reproducing kernels and the corresponding Hilbert spaces, known as reproducing kernel Hilbert spaces (RKHS), on the complex plane. This section examines the operations performed on replicating kernels and discusses important theorems related to the Bergman kernel for various domains.

REFERENCES

1. Conway JB. A Course in Functional Analysis. Berlin-Heidelberg, New York: Springer-Verlag; 1989
2. Douglas RG. Banach Algebra Techniques in Operator Theory. New York: Springer-Verlag, Academic Press; 1972
3. Okutmustur B, Gheondea A. Reproducing Kernel Hilbert Spaces: The Basics, Bergman Spaces, and Interpolation Problems of Reproducing Kernels and its Applications. Riga-Latvia: Lap Lambert Academic Publishing; 2010
4. Ando T. Reproducing Kernel Spaces and Quadratic Inequalities, Lecture Notes. Sapporo, Japan: Hokkaido University, Research Institute of Applied Electricity, Division of Applied Mathematics; 1987
5. Aronszajn N. Theory of reproducing kernels. Transactions of the American Mathematical Society. 1950;68:337-404
6. Saitoh S, Sawano Y. Theory of Reproducing Kernels and Applications. Singapore: Springer; 2016
7. Duren PL. Theory of H_p Spaces. New York: Academic Press, Inc.; 1970
8. Duren PL, Schuster A. Bergman Spaces. Providence, R.I.: American Mathematical Society; 2004
9. Koosis P. Introduction to H_p Spaces. Cambridge: Cambridge Mathematical Press; 1970

06 APPLICATIONS OF ANALYSIS IN CERTAIN HILBERT SPACES

1. INTRODUCTION: AN OVERVIEW

Hilbert spaces have a crucial role in the fields of partial differential equations, quantum physics, Fourier analysis, and ergodicity. In addition to the traditional Euclidean spaces, Hilbert spaces encompass various other types of spaces such as spaces of square-integrable functions, spaces of sequences, Sobolev spaces of generalized functions, and Hardy spaces of holomorphic functions. Saitoh et al. utilized the principles of Hilbert spaces to address the Tikhonov regularization difficulties [1, 2]. Matsuura et al. derived approximation solutions for bounded linear operator equations using numerical methods implemented on computers [3, 4]. In recent years, there has been significant interest in the theory of Hilbert spaces across several branches of mathematical sciences [5, 6, 7, 8, 9]. The findings of this study are anticipated to be valuable in the discourse surrounding uncertainty inequality for Hilbert space H and the minimizer function linked to a bounded linear operator T from H to a Hilbert space K. We consider Hardy and Dirichlet spaces as applications.

Let $\mathbb{C}$ be the complex plane and $\mathbb{D}\mathbb{D}=\{z\in\mathbb{C}:|z|<1\}$ the open unit disk. The Hardy space $H(\mathbb{D}\mathbb{D})$ is the set of all analytic functions f in the unit disk $\mathbb{D}\mathbb{D}$ with the finite integral:

$$\int_0^{2\pi} \left|f\left(e^{i\theta}\right)\right|^2 \mathrm{d}\theta. \tag{1}$$

It is a Hilbert space when equipped with the inner product:

$$\langle f, g\rangle_{H(\mathbb{D})} = \frac{1}{2\pi}\int_0^{2\pi} f\left(e^{i\theta}\right)\overline{g(e^{i\theta})}\,\mathrm{d}\theta. \tag{2}$$

Over the years, the applications of Hardy space $H(\mathbb{D}\mathbb{D})$ play an important role in various fields of mathematics [5, 10] and in certain parts of quantum mechanics [11, 12]. And this space is the background of some applications. For example, in Section 3, we study on $H(\mathbb{D}\mathbb{D})$ the following two operators:

$$\nabla f(z) = f'(z), \quad Lf(z) = z^2 f'(z) + zf(z), \tag{3}$$

and we deduce uncertainty inequality for this space. Next, we establish the minimizer function associated with the difference operator:

$$T_1 f(z) = \frac{1}{z}(f(z) - f(0)). \quad (4)$$

In Section 4, we consider the Dirichlet space $\mathcal{D}(\mathbb{D}\mathbb{D})$, which is the set of all analytic functions f in the unit disk $\mathbb{D}\mathbb{D}$ with the finite Dirichlet integral:

$$\int_{\mathbb{D}} | f'(z)|^2 \frac{dx\,dy}{\pi}, \quad z = x + iy. \quad (5)$$

It is also a Hilbert space when equipped with the inner product:

$$\langle f,g\rangle_{\mathcal{D}(\mathbb{D})} = f(0)\overline{g(0)} + \int_{\mathbb{D}} f'(z)\overline{g'(z)} \frac{dx\,dy}{\pi}, \quad z = x + iy. \quad (6)$$

This space is the objective of many applicable works [5, 13, 14, 15, 16, 17] and plays a background to our contribution. For example, we study on $\mathcal{D}(\mathbb{D}\mathbb{D})$ the following two operators:

$$\Lambda f(z) = f'(z) - f'(0), \quad Xf(z) = z^2 f'(z), \quad (7)$$

and we deduce the uncertainty inequality for this space $\mathcal{D}(\mathbb{D}\mathbb{D})$. And we establish the minimizer function associated with the difference operator:

$$T_2 f(z) = \frac{1}{z}(f(z) - zf'(0) - f(0)). \quad (8)$$

2. GENERALIZED RESULTS

Let H be a Hilbert space equipped with the inner product $\langle .,. \rangle_H$. And let A and B be the two operators defined on H. We define the commutator $[A,B]$ by

$$[A, B] := AB - BA. \quad (9)$$

The adjoint of A denoted by A^* is defined by

$$\langle Af,g\rangle_H = \langle f, A^* g\rangle_H, \quad (10)$$

for $f \in \text{Dom}(A)$ and $g \in \text{Dom}(A^*)$.

Theorem 2.1. For $f \in \text{Dom}(AA^*) \cap \text{Dom}(A^*A)$, one has

$$\|A^* f\|_H^2 = \|Af\|_H^2 + \langle [A, A^*] f,f\rangle_H. \quad (11)$$

Proof. Let $f \in \text{Dom}(AA^*) \cap \text{Dom}(A^*A)$. Then AA^*f and A^*Af belong to H. Therefore $[A,A^*] f \in H$. Hence one has

$$\|A^* f\|_H^2 = \langle AA^* f,f\rangle_H = \langle A^* Af,f\rangle_H + \langle [A,A^*]f,f\rangle_H \quad (12)$$

$$= \|Af\|_H^2 + \langle [A,A^*]f,f\rangle_H. \quad \square \quad (13)$$

The following result is proved in [18, 19].

Theorem 2.2. Let A and B be the self-adjoint operators on a Hilbert space H. Then

$$\|(A-a)f\|_H\|(B-b)f\|_H \geq \frac{1}{2}|\langle [A,B]f,f\rangle_H|, \tag{14}$$

for all $f\in$ Dom(AB) ∩Dom(BA), and all $a,b\in\mathbb{R}.\mathbb{R}$.

Theorem 2.3. Let $f\in$ Dom(AA^*) ∩Dom(A^*A). For all $a,b\in\mathbb{R}\mathbb{R}$, one has

$$\|(A+A^*-a)f\|_H\|(A-A^*+ib)f\|_H \geq |\|Af\|_H^2 - \|A^*f\|_H^2|, \tag{15}$$

where i is the imaginary unit.

Proof. Let us consider the following two operators on Dom(AA^*) ∩Dom(A^*A) Dom by

$$P = A + A^*, \quad Q = i(A - A^*). \tag{16}$$

It follows that, for $f\in$ Dom(AA^*) ∩Dom(A^*A), we have Pf, $Qf \in H$. The operators P and Q are self-adjoint and $[P,Q]$=-2i $[A,A^*]$. Thus the inequality (15) follows from Theorems 2.1 and 2.2.

Theorem 2.4. Let $f\in$Dom(AA^*) ∩Dom(A^*A). Then

$$\Delta_H^+(f)\Delta_H^-(f) \geq \|f\|_H^4\left(\|Af\|_H^2 - \|A^*f\|_H^2\right)^2, \tag{17}$$

where

$$\Delta_H^{\pm}(f) = \|f\|_H^2\|(A\pm A^*)f\|_H^2 - |\langle (A\pm A^*)f,f\rangle_H|^2. \tag{18}$$

Proof. Let $f\in$ Dom(AA^*) ∩ Dom(A^*A). The operator P given by (16) is self-adjoint; then for any real a, we have

$$\|(P-a)f\|_H^2 = \|Pf\|_H^2 + a^2\|f\|_H^2 - 2a\langle Pf,f\rangle_H. \tag{19}$$

This shows that

$$\min_{a\in\mathbb{R}}\|(P-a)f\|_H^2 = \|Pf\|_H^2 - \frac{|\langle Pf,f\rangle_H|^2}{\|f\|_H^2}, \tag{20}$$

and the minimum is attained when $a = \frac{\langle Pf,f\rangle_H}{\|f\|_H^2}$. In other words, we have

$$\min_{a\in\mathbb{R}}\|(A+A^*-a)f\|_H^2 = \|(A+A^*)f\|_H^2 - \frac{|\langle (A+A^*)f,f\rangle_H|^2}{\|f\|_H^2}. \tag{21}$$

Similarly

$$\min_{b\in\mathbb{R}}\|(A-A^*+ib)f\|_H^2 = \|(A-A^*)f\|_H^2 - \frac{|\langle (A-A^*)f,f\rangle_H|^2}{\|f\|_H^2}. \tag{22}$$

Then by (15), (21), and (22), we deduce the inequality (17).

Let λ>0 and let $T{:}H\to K$ be a bounded linear operator from H into a Hilbert space K. Building on the ideas of Saitoh [2], we examine the minimizer function associated with the operator T.

Theorem 2.5. For any $k\in K$ and for any λ>0, the problem

$$\inf_{f\in H}\{\lambda\|f\|_H^2+\|Tf-k\|_K^2\} \tag{23}$$

has a unique minimizer given by

$$f_{\lambda,k}^*=(\lambda I+T^*T)^{-1}T^*k. \tag{24}$$

Proof. The problem (23) is solved elementarily by finding the roots of the first derivative $D\Phi$ of the quadratic and strictly convex function $\Phi(f)=\lambda\|f\|_H^2+\|Tf-k\|_K^2$. Note that for convex functions, the equation $D\Phi(f)=0$ is a necessary and sufficient condition for the minimum at f. The calculation provides

$$D\Phi(f)=2\lambda f+2T^*(Tf-k), \tag{25}$$

and the assertion of the theorem follows at once.

Theorem 2.6. If $T{:}H\to K$ is an isometric isomorphism; then for any $k\in K$ and for any λ>0, the problem

$$\inf_{f\in H}\{\lambda\|f\|_H^2+\|Tf-k\|_K^2\} \tag{26}$$

has a unique minimizer given by

$$f_{\lambda,h}^*=\frac{1}{\lambda+1}T^{-1}k. \tag{27}$$

Proof. We have $T^*=T^{-1}$ and $T^*T=I$. Thus, by (24), we deduce the result.

3. THE HARDY SPACE ($\mathbb{D}\mathbb{D}$)

Let $\mathbb{C}$ be the complex plane and $\mathbb{D}\mathbb{D}=\{z\in\mathbb{C}:|z|<1\}$ the open unit disk. The Hardy space $H(\mathbb{D}\mathbb{D})$ is the set of all analytic functions f in the unit disk $\mathbb{D}\mathbb{D}$ with the finite integral:

$$\int_0^{2\pi}\left|f\left(e^{i\theta}\right)\right|^2 d\theta. \tag{28}$$

It is a Hilbert space when equipped with the inner product:

$$\langle f,g\rangle_{H(\mathbb{D})}=\frac{1}{2\pi}\int_0^{2\pi}f(e^{i\theta})\overline{g(e^{i\theta})}\,d\theta. \tag{29}$$

If $f,g\in H(\mathbb{D}\mathbb{D})$ with $f(z)=\sum_{n=0}^{\infty}a_nz^n$ and $g(z)=\sum_{n=0}^{\infty}b_nz^n$, then

$$\langle f,g\rangle_{H(\mathbb{D})}=\sum_{n=0}^{\infty}a_n\overline{b_n}. \tag{30}$$

The set $\{z^n\}_{n=0}^{\infty}$ forms an Hilbert's basis for the space $H(\mathbb{D}\mathbb{D})$.
The Szegő kernel S_z given for $z \in \mathbb{D}\mathbb{D}$, by

$$S_z(w) = \sum_{n=0}^{\infty} \bar{z}^n w^n = \frac{1}{1 - \bar{z}w}, \quad w \in \mathbb{D}, \tag{31}$$

is a reproducing kernel for the Hardy space $H(\mathbb{D}\mathbb{D})$, meaning that $S_z \in H(\mathbb{D}\mathbb{D})$, and for all $f \in H(\mathbb{D}\mathbb{D})$, we have $\langle f, S_z \rangle_{H(\mathbb{D}\mathbb{D})} = f(z)$.

For $z \in \mathbb{D}\mathbb{D}$, the function $u(z) = S_{\bar{z}}(w)$ is the unique analytic solution on $\mathbb{D}\mathbb{D}$ of the initial problem:

$$u'(z) = w(zu'(z) + u(z)), \quad w \in \mathbb{D}, \quad u(0) = 1. \tag{32}$$

In the next of this section, we define the operators ∇, $\mathfrak{R}$, and L on $H(\mathbb{D}\mathbb{D})$ by

$$\nabla f(z) = f'(z), \quad \mathfrak{R}f(z) = zf'(z), \quad Lf(z) = z^2 f'(z) + zf(z). \tag{33}$$

These operators satisfy the commutation rule:

$$[\nabla, L] = \nabla L - L\nabla = 2\mathfrak{R} + I, \tag{34}$$

where I is the identity operator.

We define the Hilbert space $U(\mathbb{D}\mathbb{D})$ as the space of all analytic functions f in the unit disk $\mathbb{D}\mathbb{D}$ such that

$$\|f\|^2_{U(\mathbb{D})} = \frac{1}{2\pi} \int_0^{2\pi} |f'(e^{i\theta})|^2 \, d\theta < \infty. \tag{35}$$

If $f \in U(\mathbb{D}\mathbb{D})$ with $f(z) = \sum_{n=0}^{\infty} a_n z^n$, then

$$\|f\|^2_{U(\mathbb{D})} = \sum_{n=1}^{\infty} n^2 |a_n|^2. \tag{36}$$

Thus, the space $U(\mathbb{D}\mathbb{D})$ is a subspace of the Hardy space $H(\mathbb{D}\mathbb{D})$.

Theorem 3.1.

(i) For $f \in U(\mathbb{D}\mathbb{D})$, then ∇f, $\mathfrak{R}f$ and Lf belong to $H(\mathbb{D}\mathbb{D})$.

(ii) $\nabla^* = L$.

(iii) For $f \in U(\mathbb{D}\mathbb{D})$, one has

$$\|Lf\|^2_{H(\mathbb{D})} = \|\nabla f\|^2_{H(\mathbb{D})} + \|f\|^2_{H(\mathbb{D})} + 2\langle \mathfrak{R}f, f \rangle_{H(\mathbb{D})}. \tag{37}$$

Proof.

(i) Let $f \in U(\mathbb{D}\mathbb{D})$ with $f(z) = \sum_{n=0}^{\infty} a_n z^n$. Then

$$\nabla f(z) = \sum_{n=0}^{\infty} (n+1) a_{n+1} z^n, \quad \mathfrak{R}f(z) = \sum_{n=1}^{\infty} n a_n z^n, \tag{38}$$

and

$$Lf(z) = \sum_{n=1}^{\infty} na_{n-1}z^n. \tag{39}$$

Therefore

$$\|\nabla f\|_{H(\mathbb{D})}^2 = \sum_{n=0}^{\infty} (n+1)^2 |a_{n+1}|^2 = \|f\|_{U(\mathbb{D})}^2, \tag{40}$$

$$\|\Re f\|_{H(\mathbb{D})}^2 = \sum_{n=1}^{\infty} n^2 |a_n|^2 = \|f\|_{U(\mathbb{D})}^2, \tag{41}$$

and

$$\|Lf\|_{H(\mathbb{D})}^2 = \sum_{n=0}^{\infty} (n+1)^2 |a_n|^2 \leq |f(0)|^2 + 4\|f\|_{U(\mathbb{D})}^2. \tag{42}$$

Consequently ∇f, $\Re f$, and Lf belong to $H(\mathbb{D}\mathbb{D})$.

(ii) For $f, g \in U(\mathbb{D})$ with $f(z) = \sum_{n=0}^{\infty} a_n z^n$ and $g(z) = \sum_{n=0}^{\infty} b_n z^n$, one has

$$\langle \nabla f, g\rangle_{H(\mathbb{D})} = \sum_{n=0}^{\infty} (n+1)a_{n+1}\overline{b_n} = \sum_{n=1}^{\infty} na_n\overline{b_{n-1}} = \langle f, Lg\rangle_{H(\mathbb{D})}. \tag{43}$$

Thus $\nabla^*=L$.

(iii) Let $f \in U(\mathbb{D}\mathbb{D})$. By (ii) and (34), we deduce that

$$\|Lf\|_{H(\mathbb{D})}^2 = \langle \nabla Lf, f\rangle_{H(\mathbb{D})} \tag{44}$$

$$= \langle L\nabla f, f\rangle_{H(\mathbb{D})} + \langle [\nabla, L]f, f\rangle_{H(\mathbb{D})} \tag{45}$$

$$= \|\nabla f\|_{H(\mathbb{D})}^2 + \|f\|_{H(\mathbb{D})}^2 + 2\langle \Re f, f\rangle_{H(\mathbb{D})}. \quad \square \tag{46}$$

Theorem 3.2. Let $f \in U(\mathbb{D}\mathbb{D})$. For all $a, b \in \mathbb{R}\mathbb{R}$, one has

$$\|(\nabla + L - a)f\|_{H(\mathbb{D})} \|(\nabla - L + ib)f\|_{H(\mathbb{D})} \geq \|f\|_{H(\mathbb{D})}^2 + 2\langle \Re f, f\rangle_{H(\mathbb{D})}. \tag{47}$$

Theorem 3.3. Let T_1 be the difference operator defined on $H(\mathbb{D}\mathbb{D})$ by

$$T_1 f(z) = \frac{1}{z}(f(z) - f(0)). \tag{48}$$

(i) The operator T_1 maps continuously from $H(\mathbb{D}\mathbb{D})$ to $H(\mathbb{D}\mathbb{D})$, and

$$\|T_1 f\|_{H(\mathbb{D})} \leq \|f\|_{H(\mathbb{D})}. \tag{49}$$

(ii) For $f \in H(\mathbb{D}\mathbb{D})$ and $z \in \mathbb{D}\mathbb{D}$, we have

$$T_1^* f(z) = zf(z), \quad T_1^* T_1 f(z) = f(z) - f(0). \tag{50}$$

(iii) For any $h \in H(\mathbb{D}\mathbb{D})$ and for any $\lambda>0$, the problem

$$\inf_{f \in H(\mathbb{D})} \left\{ \lambda \|f\|^2_{H(\mathbb{D})} + \|T_1 f - h\|^2_{H(\mathbb{D})} \right\} \quad (51)$$

has a unique minimizer given by

$$f^*_{\lambda,h}(z) = \frac{1}{\lambda+1} zh(z), \quad z \in \mathbb{D}. \quad (52)$$

Proof.

(i) If $f \in H(\mathbb{D}\mathbb{D})$ with $f(z) = \sum_{n=0}^{\infty} a_n z^n$, then $T_1 f(z) = \sum_{n=0}^{\infty} a_{n+1} z^n$ and

$$\|T_1 f\|^2_{H(\mathbb{D})} = \sum_{n=1}^{\infty} |a_n|^2 \leq \|f\|^2_{H(\mathbb{D})}. \quad (53)$$

(ii) If $f, g \in H(\mathbb{D}\mathbb{D})$ with $f(z) = \sum_{n=0}^{\infty} a_n z^n$ and $g(z) = \sum_{n=0}^{\infty} b_n z^n$, then

$$\langle T_1 f, g \rangle_{H(\mathbb{D})} = \sum_{n=0}^{\infty} a_{n+1} \overline{b_n} = \sum_{n=1}^{\infty} a_n \overline{b_{n-1}} = \langle f, T_1^* g \rangle_{H(\mathbb{D})}, \quad (54)$$

where $T_1^* g(z)$, for $z \in \mathbb{D}$. And therefore

$$T_1^* T_1 f(z) = z T_1 f(z) = f(z) - f(0). \quad (55)$$

(iii) From Theorem 2.5 we have

$$(\lambda I + T_1^* T_1) f^*_{\lambda,h}(z) = T_1^* h(z). \quad (56)$$

By (ii) we deduce that

$$(\lambda+1) f^*_{\lambda,h}(z) - f^*_{\lambda,h}(0) = zh(z). \quad (57)$$

And from this equation, $f^*_{\lambda,h}(0) = 0$. Hence

$$f^*_{\lambda,h}(z) = \frac{1}{\lambda+1} zh(z). \quad \square \quad (58)$$

4. THE DIRICHLET SPACE D ($\mathbb{D}$)

The Dirichlet space $\mathcal{D}(\mathbb{D})$ is the set of all analytic functions f in the unit disk $\mathbb{D}$ with the finite Dirichlet integral:

$$\int_{\mathbb{D}} |f'(z)|^2 \frac{dx\,dy}{\pi}, \quad z = x + iy. \quad (59)$$

It is a Hilbert space when equipped with the inner product:

$$\langle f, g \rangle_{\mathcal{D}(\mathbb{D})} = f(0)\overline{g(0)} + \int_{\mathbb{D}} f'(z)\overline{g'(z)} \frac{dx\,dy}{\pi}, \quad z = x + iy. \quad (60)$$

If $f, g \in \mathcal{D}(\mathbb{D})$ with $f(z) = \sum_{n=0}^{\infty} a_n z^n$ and $g(z) = \sum_{n=0}^{\infty} b_n z^n$, then

$$\langle f,g\rangle_{\mathcal{D}(\mathbb{D})} = a_0\overline{b_0} + \sum_{n=1}^{\infty} n a_n \overline{b_n}. \tag{61}$$

The set $\left\{1, \frac{z^n}{\sqrt{n}}\right\}_{n=1}^{\infty}$ forms an Hilbert's basis for the space $\mathcal{D}(\mathbb{D}\mathbb{D})$.

The function K_z given for z∈$\mathbb{D}\mathbb{D}$, by

$$K_z(w) = 1 + \log\left(\frac{1}{1-\bar{z}w}\right), \quad w \in \mathbb{D}, \tag{62}$$

is a reproducing kernel for the Dirichlet space $\mathcal{D}(\mathbb{D}\mathbb{D})$, meaning that $K_z \in \mathbb{D}(\mathbb{D}\mathbb{D})$, and for all $f \in \mathcal{D}(\mathbb{D}\mathbb{D})$, we have $\langle f,K_z\rangle_{\mathcal{D}(\mathbb{D})} = f(z)$.

For z∈$\mathbb{D}\mathbb{D}$, the function u(z)= $K_{\bar{z}}(w)$ is the unique analytic solution on $\mathbb{D}\mathbb{D}$ of the initial problem:

$$\frac{u'(z) - u'(0)}{z} = wu'(z), \quad w \in \mathbb{D}, \quad u(0) = 1. \tag{63}$$

In the next of this section, we define the operators Λ, $\mathfrak{R}$, and X on $\mathcal{D}(\mathbb{D}\mathbb{D})$ by

$$\Lambda f(z) = f'(z) - f'(0), \quad \mathfrak{R}f(z) = zf'(z), \quad Xf(z) = z^2 f'(z). \tag{64}$$

These operators satisfy the following commutation relation:

$$[\Lambda, X] = \Lambda X - X\Lambda = 2\mathfrak{R}. \tag{65}$$

We define the Hilbert space $V(\mathbb{D}\mathbb{D})$ as the space of all analytic functions f in the unit disk $\mathbb{D}\mathbb{D}$ such that

$$\|f\|^2_{V(\mathbb{D})} = \int_{\mathbb{D}} |f'(z)|^2 |z|^2 \frac{dx\,dy}{\pi} < \infty, \quad z = x + iy. \tag{66}$$

If $f \in \mathcal{V}(\mathbb{D}\mathbb{D})$ with $f(z) = \sum_{n=0}^{\infty} a_n z^n$, then

$$\|f\|^2_{V(\mathbb{D})} = \sum_{n=1}^{\infty} n^3 |a_n|^2. \tag{67}$$

Thus, the space $V(\mathbb{D}\mathbb{D})$ is a subspace of the Dirichlet space $\mathcal{D}(\mathbb{D}\mathbb{D})$.

Theorem 4.1.

(i) For $f \in V(\mathbb{D}\mathbb{D})$, then Λf, $\mathfrak{R}f$, and Xf belong to $\mathcal{D}(\mathbb{D}\mathbb{D})$.

(ii) Λ^*=X.

(iii) For $f \in V(\mathbb{D}\mathbb{D})$, one has

$$\|Xf\|^2_{\mathcal{D}(\mathbb{D})} = \|\Lambda f\|^2_{\mathcal{D}(\mathbb{D})} + 2\langle \mathfrak{R}f, f\rangle_{\mathcal{D}(\mathbb{D})}. \tag{68}$$

Proof.

(i) Let $f \in V(\mathbb{D})$ with $f(z) = \sum_{n=0}^{\infty} a_n z^n$. Then

$$\Lambda f(z) = \sum_{n=1}^{\infty} (n+1) a_{n+1} z^n, \quad \mathfrak{R} f(z) = \sum_{n=1}^{\infty} n a_n z^n, \tag{69}$$

and

$$X f(z) = \sum_{n=2}^{\infty} (n-1) a_{n-1} z^n. \tag{70}$$

Therefore

$$\|\Lambda f\|^2_{\mathcal{D}(\mathbb{D})} = \sum_{n=1}^{\infty} n(n+1)^2 |a_{n+1}|^2 \leq \sum_{n=2}^{\infty} n^3 |a_n|^2 \leq \|f\|^2_{V(\mathbb{D})}, \tag{71}$$

$$\|\mathfrak{R} f\|^2_{\mathcal{D}(\mathbb{D})} = \sum_{n=1}^{\infty} n^3 |a_n|^2 = \|f\|^2_{V(\mathbb{D})}, \tag{72}$$

and

$$\|X f\|^2_{\mathcal{D}(\mathbb{D})} = \sum_{n=1}^{\infty} (n+1) n^2 |a_n|^2 \leq 2\|f\|^2_{V(\mathbb{D})}. \tag{73}$$

Consequently Λf, $\mathfrak{R} f$, and $X f$ belong to $\mathcal{D}(\mathbb{D})$.

(ii) For $f, g \in V(\mathbb{D})$ with $f(z) = \sum_{n=0}^{\infty} a_n z^n$ and $g(z) = \sum_{n=0}^{\infty} b_n z^n$, one has

$$\langle \Lambda f, g\rangle_{\mathcal{D}(\mathbb{D})} = \sum_{n=1}^{\infty} n(n+1) a_{n+1} \overline{b_n} = \sum_{n=2}^{\infty} n(n-1) a_n \overline{b_{n-1}} = \langle f, Xg\rangle_{\mathcal{D}(\mathbb{D})}. \tag{74}$$

(iii) Let $f \in V(\mathbb{D})$. By (ii) and (65), we deduce that

$$\|X f\|^2_{\mathcal{D}(\mathbb{D})} = \langle \Lambda X f, f\rangle_{\mathcal{D}(\mathbb{D})} \tag{75}$$

$$= \langle X \Lambda f, f\rangle_{\mathcal{D}(\mathbb{D})} + \langle [\Lambda, X] f, f\rangle_{\mathcal{D}(\mathbb{D})} \tag{76}$$

$$= \|\Lambda f\|^2_{\mathcal{D}(\mathbb{D})} + 2\langle \mathfrak{R} f, f\rangle_{\mathcal{D}(\mathbb{D})}. \quad \square \tag{77}$$

Theorem 4.2. Let $f \in V(\mathbb{D})$. For all $a, b \in \mathbb{R}$, one has

$$\|(\Lambda + X - a) f\|_{\mathcal{D}(\mathbb{D})} \|(\Lambda - X + ib) f\|_{\mathcal{D}(\mathbb{D})} \geq 2\langle \mathfrak{R} f, f\rangle_{\mathcal{D}(\mathbb{D})}. \tag{78}$$

Theorem 4.3. Let T_2 be the difference operator defined on $\mathcal{D}(\mathbb{D})$ by

$$T_2 f(z) = \frac{1}{z}\left(f(z) - z f'(0) - f(0)\right). \tag{79}$$

(i) The operator T_2 maps continuously from $\mathcal{D}(\mathbb{D}\mathbb{D})$ to $\mathcal{D}(\mathbb{D}\mathbb{D})$, and

$$\|T_2 f\|_{\mathcal{D}(\mathbb{D})} \leq \|f\|_{\mathcal{D}(\mathbb{D})}. \tag{80}$$

(ii) For $f \in \mathcal{D}(\mathbb{D}\mathbb{D})$ with $f(z) = \sum_{n=0}^{\infty} a_n z^n$, we have

$$T_2^* f(z) = \sum_{n=2}^{\infty} \frac{n-1}{n} a_{n-1} z^n, \quad T_2^* T_2 f(z) = \sum_{n=2}^{\infty} \frac{n-1}{n} a_n z^n. \tag{81}$$

(iii) For any $d \in \mathcal{D}(\mathbb{D}\mathbb{D})$ and for any $\lambda > 0$, the problem

$$\inf_{f \in \mathcal{D}(\mathbb{D})} \left\{ \lambda \|f\|_{\mathcal{D}(\mathbb{D})}^2 + \|T_2 f - d\|_{\mathcal{D}(\mathbb{D})}^2 \right\} \tag{82}$$

has a unique minimizer given by

$$f_{\lambda,d}^*(z) = \langle d, \Psi_z \rangle_{\mathcal{D}(\mathbb{D})}, \quad z \in \mathbb{D}, \tag{83}$$

$$\Psi_z(w) = \sum_{n=1}^{\infty} \frac{\overline{z}^{n+1}}{\lambda(n+1) + n} w^n, \quad w \in \mathbb{D}. \tag{84}$$

Proof.

(i) If $f \in \mathcal{D}(\mathbb{D}\mathbb{D})$ with $f(z) = \sum_{n=0}^{\infty} a_n z^n$, then $T_2 f(z) = \sum_{n=1}^{\infty} a_{n+1} z^n$ and

$$\|T_2 f\|_{\mathcal{D}(\mathbb{D})}^2 = \sum_{n=2}^{\infty} (n-1)|a_n|^2 \leq \sum_{n=2}^{\infty} n|a_n|^2 \leq \|f\|_{\mathcal{D}(\mathbb{D})}^2. \tag{85}$$

(ii) If $f, g \in \mathcal{D}(\mathbb{D}\mathbb{D})$ with $f(z) = \sum_{n=0}^{\infty} a_n z^n$ and $g(z) = \sum_{n=0}^{\infty} b_n z^n$, then

$$\langle T_2 f, g \rangle_{\mathcal{D}(\mathbb{D})} = \sum_{n=1}^{\infty} n a_{n+1} \overline{b_n} = \sum_{n=2}^{\infty} (n-1) a_n \overline{b_{n-1}} = \langle f, T_2^* g \rangle_{\mathcal{D}(\mathbb{D})}, \tag{86}$$

where

$$T_2^* g(z) = \sum_{n=2}^{\infty} \frac{n-1}{n} b_{n-1} z^n, \quad z \in \mathbb{D}. \tag{87}$$

And therefore

$$T_2^* T_2 f(z) = \sum_{n=2}^{\infty} \frac{n-1}{n} a_n z^n. \tag{88}$$

(i) We put $d(z) = \sum_{n=0}^{\infty} d_n z^n$ and

$$f_{\lambda,d}^*(z) = \sum_{n=0}^{\infty} c_n z^n. \tag{89}$$

From (ii) and the equation

$$(\lambda I + T_2^* T_2) f_{\lambda,d}^*(z) = T_2^* d(z), \tag{90}$$

we deduce that

$$c_1 = c_0 = 0, \quad c_n = \frac{n-1}{\lambda n + n - 1} d_{n-1}, \quad n \geq 2. \tag{91}$$

Thus

$$f^*_{\lambda,d}(z) = \sum_{n=1}^{\infty} \frac{n d_n}{\lambda(n+1) + n} z^{n+1} = \langle d, \Psi_z \rangle_{\mathcal{D}(\mathbb{D})}, \quad z \in \mathbb{D}. \quad \square \tag{92}$$

AT A GLANCE

This work presents an uncertainty inequality for a Hilbert space H. The minimizer function for a bounded linear operator from the Hilbert space H to the Hilbert space K is given. We have obtained certain findings concerning Hardy and Dirichlet spaces on the unit disk D.

REFERENCES

1. Saitoh S. The Weierstrass transform and an isometry in the heat equation. Applicable Analysis. 1983;16:1-6
2. Saitoh S. Best approximation, Tikhonov regularization and reproducing kernels. Kodai Mathematical Journal. 2005;28:359-367
3. Matsuura T, Saitoh S, Trong DD. Inversion formulas in heat conduction multidimensional spaces. Journal of Inverse and Ill-Posed Problems. 2005;13:479-493
4. T, Saitoh S. Analytical and numerical inversion formulas in the Gaussian convolution by using the Paley-Wiener spaces. Applicable Analysis. 2006;85:901-915
5. Paulsen VI. An Introduction to the Theory of Reproducing Kernel Hilbert Spaces. Cambridge: Cambridge University Press; 2016
6. Saitoh S. Approximate real inversion formulas of the Gaussian convolution. Applicable Analysis. 2004;83:727-733
7. F. Operators and Tikhonov regularization on the Fock space. Integral Transforms and Special Functions. 2014;25(4):283-294
8. Soltani F. Inversion formulas for the Dunkl-type Segal-Bargmann transform. Integral Transforms and Special Functions. 2015;26(5):325-339
9. Soltani F. Dunkl multiplier operators on a class of reproducing kernel Hilbert spaces. Journal of Mathematical Research with Applications. 2016;36(6):689-702
10. Tuan VK, Hong NT. Interpolation in the Hardy space. Integral Transforms and Special Functions. 2013;24(8):664-671

11. Bohm A, Bui HV. The marvelous consequences of Hardy spaces in quantum physics. Geometric Methods in Physics. 2013;30(1):211-228
12. Mouayn Z. Resolution of the identity of the classical Hardy space by means of Barut-Girardello coherent states. Mathematical Physics. 2012;2012: Article ID 530473, 12 p
13. N, Rochberg R, Sawyer ET, Wick BD. The Dirichlet space: A survey. New York Journal of Mathematics. 2011;17a:45-86
14. R. Toeplitz operators on Dirichlet-type spaces, J. Operator Theory. 2002;48(1):3-13
15. L, Tong C, Zeng H. Some linear isometric operators on the Dirichlet space. Applied Mathematics & Information Sciences. 2012;6(1):265-270
16. LG, Zhou ZH, Dong XT. Isometric composition operators on weighted Dirichlet-type spaces. Journal of Inequalities and Applications. 2012;2012:23
17. MJ, Vukotic D. Isometries of the Dirichlet space among the composition operators. Proceedings of the American Mathematical Society. 2005;134:1701-1705
18. Folland G. Harmonic analysis on phase space. In: Annals of Mathematics Studies. Vol. 122. Princeton, New Jersey: Princeton University Press; 1989
19. Gröchenig K. Foundations of Time-frequency Analysis. Boston: Birkhäuser; 2001

07 ANALYZING THE FLUCTUATIONS OF PM10 IN THE HILBERT SPACE BY SPECTRAL OBSERVATIONS

1. INTRODUCTION: AN OVERVIEW

The concentration of air pollutants typically fluctuates and is influenced by local pollutant emissions, as well as meteorological and topographical factors [1, 2]. Particulate matter (PM) consists of a diverse combination of elemental and organic carbon, ammonium, nitrates, sulfates, mineral dust, trace elements, and water [3]. Particulate matter (PM) with an aerodynamic diameter less than 10 μm, referred to as PM10, is widely recognized for its detrimental effects on human health [4]. Several studies have demonstrated that exposure to particulate matter (PM) leads to an increase in hospital admissions for cardiovascular illness, acute bronchitis, asthma episodes, respiratory disease, and congestive heart failure [5, 6, 7, 8]. In the Caribbean region, African dust is a significant contributor to the emission of PM10, which is mostly generated by large-scale sources [9]. Understanding the mechanics of the PM10 process is essential for developing strategies and creating tools to forecast dust outbreaks. The time-frequency distribution of a signal reveals how the spectral characteristics of the signal change over time. This makes it a valuable tool for examining, studying, and understanding signals that are not constant over time [10]. The necessity of a time-frequency representation (TFR) arises from the insufficiency of traditional approaches in either the time domain or frequency domain to completely characterize the properties of nonstationary signals [10]. Within the realm of literature, there are several techniques for determining the energy density in relation to both time and frequency concurrently. These techniques include the short-time Fourier transform (STFT), Hilbert-Huang transform (HHT), and wavelet transform (WT) [10, 11, 12].

This study initially examines the scaling features of PM10 data, and then proceeds to investigate the TFR. To emphasize the effectiveness of the Hilbert space, an examination of PM10 data was also conducted in the Fourier domain. The structure of this chapter is as follows. Section 2 provides an analysis of the PM10 data conducted in this investigation. Section 3 outlines the techniques employed to study the dynamics of PM10. Section 4 provides an analysis of the collected results and thereafter engages in a discussion of these findings.

2. EXPERIMENTAL DATA

Guadeloupe archipelago is a French West Indies island located in the middle of the Caribbean basin, i.e., 16.25°N latitude and 61.58°W longitude, which experiences a tropical and humid climate [13, 14]. The time series analyzed here belong to Guadeloupe air quality network which is managed by the Gwad'Air agency (http://www.gwadair.fr/). PM10 concentrations are measured at Pointe-à-Pitre (16.2422°N 61.5414°W) using the Thermo Scientific tapered element oscillating microbalance (TEOM) models 1400ab and 1400-FDMS. Hourly PM10 concentrations were sampled during the period from 1 January 2005 to 31 December 2010. We processed these data into daily average concentrations. In total, there are 2150 daily averaged data points available continuously for 6 years. Figure 1 displays PM10 daily signal illustrating huge fluctuations and thus indicating a strong variability. These strong oscillations observed in the middle of each year are attributed to PM10 related to dust outbreaks coming from the African coast from May to September [9]. For the rest of the year, PM10 is mainly generated by anthropogenic pollution [15].

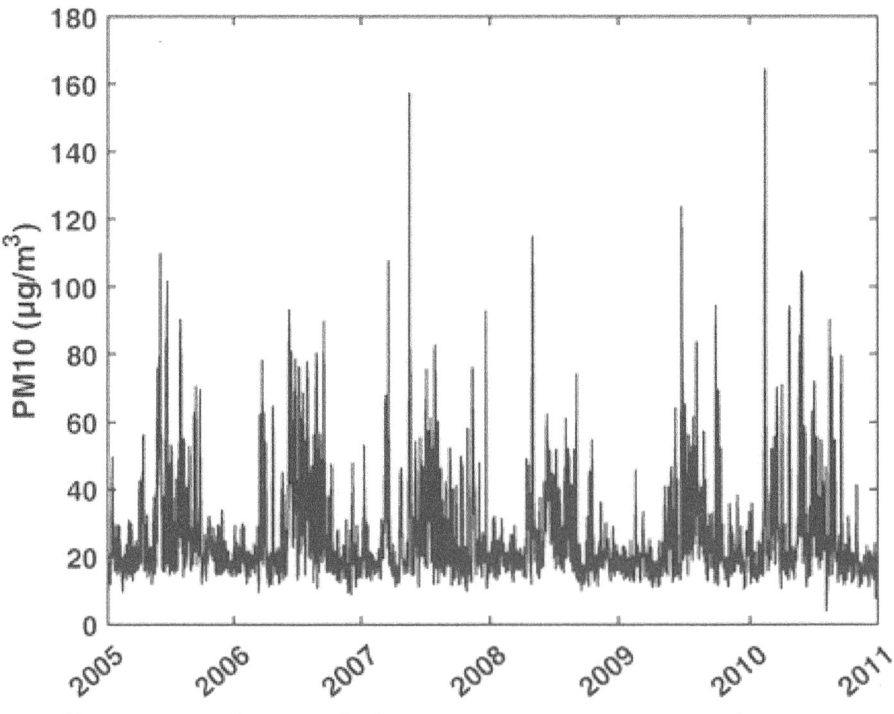

Figure 1. Illustration of PM10 daily average concentrations between 2005 and 2010, highlighting intermittent burst events with huge fluctuations.

3. METHODS

3.1 Scaling analysis (1D representation)

The description of natural phenomena by the study of statistical scale laws is not recent [16]. Self-similarity of complex systems has been widely observed in nature and is the simplest form of scale invariance. A scale invariance can be detected by computing of power spectral density (PSD). The PSD separates and measures the amount of variability occurring in different frequency bands. In this study, PSD are estimated through the Fourier and Hilbert spaces.

3.1.1 Fourier analysis

In order to investigate the scaling properties of PM10 data, classically the discrete Fourier transform of the times series considered is computed. The expression of Fourier transform *X(f)* for a process *x(t)* is recalled here. An *N* point-long interval is used to construct the value at frequency domain point *f*, X_f [17]:

$$X(f) = \int_{-T}^{+T} x(t)e^{-2\pi ift}\mathrm{dt} \tag{1}$$

Thus, the analytical expression of *X(f)* is [18]

$$|X(f)| = \sqrt{\mathcal{R}e^2(X(f)) + \mathcal{I}m^2(X(f))} \tag{2}$$

Consequently the power spectral density *E(f)* is estimated by computing the following expression:

$$E(f) = |X(f)|^2 \tag{3}$$

3.1.2 Hilbert analysis

To determine the scale invariance of a given time series in a joint amplitude-frequency space, the Hilbert-Huang transform [19, 20] is performed. HHT can be summarized in two steps: (i) empirical mode decomposition (EMD) and (ii) Hilbert spectral analysis (HSA). Empirical mode decomposition is a powerful tool to separate a nonlinear and nonstationary time series into a sum of intrinsic mode functions (IMF) without a priori basis as required by traditional Fourier-based method [19, 20, 21]. An IMF must satisfy the following two conditions: (i) the difference between the number of local extrema and the number of zero-crossings must be zero or one, and (ii) the local maxima and the envelope defined by the local minima are close to zero. Therefore, the original signal *x*(*t*) is decomposed into a sum of *n*1 IMF modes with the residual $r_n(t)$:

$$x(t) = \sum_{m=1}^{n-1} C_m(t) + r_n(t) \tag{4}$$

To obtain a physically significant IMF, this selection process must be stopped by a certain criterion. For more details, EMD decomposition is widely described in the literature [19, 20, 21, 22, 23].

To characterize the time-frequency energy distribution from the original signal $x(t)$, HSA is applied on each obtained IMF component $C_m(t)$ to extract the instantaneous amplitude and frequency [19, 24]. The Hilbert transform is defined by:

$$\tilde{C}_m(t) = \frac{1}{\pi} P \int_{-\infty}^{+\infty} \frac{C_m(t')}{t - t'} dt' \tag{5}$$

with P the Cauchy principal value [24, 25]. We can specify an analytical signal z for each IMF mode $C_m(t)$ with

$$Z_m(t) = C_m(t) + j\tilde{C}_m(t) = A_m(t)e^{j\varphi_m(t)} \tag{6}$$

where $A_m(t) = |Z_m(t)| = \sqrt{C_m(t)^2 + \tilde{C}_m(t)^2}$ describes an amplitude and $\varphi_m(t) = \arg(z) = \arctan\left[\frac{\tilde{C}_m(t)}{C_m(t)}\right]$ represents the phase function of IMF modes. Consequently, the instantaneous frequency $\omega_m(t)$ is defined from the phase $\varphi_m(t)$ by

$$\omega_m(t) = \frac{1}{2\pi} \frac{d\varphi_m(t)}{dt} \tag{7}$$

Thus, the original signal $x(t)$ can be expressed as

$$x(t) = \mathcal{R}e \sum_{m=1}^{N} A_m(t)e^{j\varphi_m(t)} = \mathcal{R}e \sum_{m=1}^{N} A_m(t)e^{j\int_{-\infty}^{t} \omega_m(t)dt} \tag{8}$$

where $\mathcal{R}_e$ is a part real [19, 20, 26].

Due to the simultaneous representation of frequency modulation and amplitude modulation, the HHT can be considered as a generalization of the Fourier transform [19, 20]. The energy in a time-frequency space is designated as the Hilbert spectrum with $H(\omega,t)=\mathcal{A}^2(\omega,t)$. The Hilbert marginal spectrum $h(\omega)$ is defined by

$$h(\omega) = \frac{1}{T}\int_0^T H(\omega,t)dt \tag{9}$$

where T is the total data length. The Hilbert spectrum $H(\omega,t)$ gives a measure of amplitude from each frequency and time, while the marginal spectrum $h(\omega)$ gives a measure of the total amplitude from each frequency [27]. As a result, the marginal spectrum can be compared to the Fourier spectrum [19, 20].

In conclusion, for a scale-invariant process, the Fourier $E(f)$ and the Hilbert $h(\omega)$ spectral densities obtained follow a power law over a range of frequencies:

$$E(f) \sim f^{-\beta_f} \tag{10}$$

$$h(\omega) \sim \omega^{-\beta_h} \tag{11}$$

where f and ω are the frequencies and β_f and β_h are the spectral exponents, respectively, in the Fourier and Hilbert spaces. It reveals the scale-free memory effect as a power law dependence of the frequency distribution. Consequently, β_f and β_h contain information about the degree of stationarity of the studied parameter [16, 28, 29]:

- If β_f or β_h <1, the process is stationary.
- If β_f or β_h >1, the process is nonstationary.
- If 1< β_f or β_h <3, the process is nonstationary with increments stationary.
- Spectral analysis has been widely applied in various research fields [30, 31, 32, 33, 34].

3.2 Time-frequency representation (2D representation)

3.2.1 Spectrogram

The spectrogram (SPEC) of a signal $x(t)$ is defined as the squared magnitudes of the STFT as shown in Eq. (12) [12]:

$$\mathrm{SPEC}_x(t,f) = |S_x(t,f)|^2 \tag{12}$$

where $S_x(t,f) = \int_{-\infty}^{+\infty} x(\tau)w(\tau - t)e^{-jf\tau}d\tau$ is the STFT of $x(t)$, $w(\tau)$ is a window (e.g., Hanning, rectangular, Hamming), t is time, and f is frequency.

As depicted in Eq. (13), SPEC roughly describes the energy density of the signal at point (t,f) [12]:

$$\int_{-\infty}^{+\infty}\int_{-\infty}^{+\infty} \mathrm{SPEC}_x(t,f)\mathrm{dtdf} = \int_{-\infty}^{+\infty} |x(t)|^2\mathrm{dt} \tag{13}$$

The SPEC has been applied successfully in various research fields [12, 35, 36, 37]. The main advantages of SPEC are an easily understanding interpretation, and it allows a fast computation. However, the main drawback of SPEC is the same as that of the STFT [12]. Indeed, there is a trade-off between time and frequency resolution.

3.2.2 Hilbert spectrum

The Hilbert spectrum (HS) is a joint time-frequency representation introduced by [19]. It is important to notice that the two important tools (i.e., EMD and HS) for exploratory analysis of the data are provided by HSA method. This approach was applied successfully in various research fields as fault diagnosis for rolling bearing [11], turbulence [38], environment [34, 39], and geophysics [40], to cite a few.

4. RESULTS

4.1 Scaling properties

In order to identify the presence of scaling in PM10 time series, the PSD is estimated in the Hilbert and Fourier spaces. Figure 2 depicts the power spectral density provided by the Hilbert transform and the Fourier transform. On this figure, we try to detect a power law behavior of the form $h(\omega)\sim\omega^{-\beta h}$ and $E(f)\sim f^{-\beta f}$ where β_h and β_f are, respectively, the spectral exponents in the Hilbert and Fourier spaces. On the frequency range $2.09\times10^{-7}\leqslant f\leqslant4.57\times10^{-5}$ Hz which corresponds to time scales 6.1hours$\leqslant$T$\leqslant$55.4 days, a power law behavior is clearly noticed in the Hilbert space with an estimated spectral exponent β_h = 1.02 ± 0.10. β_h is equal to 1 power law scaling observed in the mesoscale range [41]. In the Fourier space, this power law is not significant. This is due to the existence of intermittent dust events with huge fluctuations in PM10 data (see Figure 1). Indeed, the Fourier transform is a linear asymptotic approach which requires high-order harmonic components to mimic nonlinear and nonstationary process [42]. Thus, the high-order harmonics may lead an artificial energy transfer flux from a large scale (low frequency) to a small scale (high frequency) in the Fourier space. Consequently, the Fourier-based spectrum may be contaminated by this artificial energy flux [42]. The artificial energy transfer may give a less steep power spectrum as we observed in Figure 2. By contrast, combined with the EMD method, HSA has very local abilities both in physical and spectral domains and does not require any higher-order harmonic components to simulate the nonlinear and nonstationary events. As a consequence, HSA method may provide a more accurate scaling exponent and singularity spectrum [42].

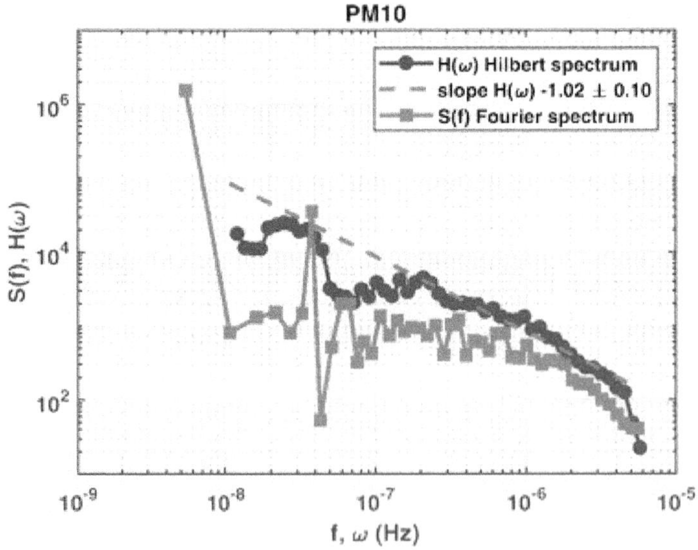

Figure 2. The spectrum of PM10 time series in the Hilbert space and the Fourier space. A power law behavior is significant only in the Hilbert space.

According to [43], wind speed dominates the amount of pollutant dispersion in the atmospheric boundary layer. In addition, this meteorological parameter could also transport PM10 from large-scale sources, i.e., African dust [9]. To complete our results, we used hourly wind speed measurements provided by the French weather office (Météo France Guadeloupe) located at Abymes (16.2630°N 61.5147°W).

PM10 and wind speed measurements are very close, i.e., ≈8.1 km of distance, and performed at the center of the island under the same atmospheric conditions [2]. Figure 3 illustrates the PSD provided by the Hilbert transform and the Fourier transform for wind speed data. This time, a power law behavior is observed in both spaces on the same frequency range $3.54\times10^{-7}\leqslant f\leqslant1.36\times10^{-4}$ Hz which corresponds to time scales 2.1hours$\leqslant T\leqslant$32.7 days. Contrary to PM10 which is a passive scalar, wind speed is a vector quantity. The estimated spectral exponents are identical with, respectively, 0.89 ± 0.06 and 0.90 ± 0.12 in the Hilbert and Fourier spaces. As PM10, spectral exponent values are also close to −1. For wind speed, at low frequencies, a spectrum close to the 1 power law is likely occurs close to a rough surface, due to a strong interaction between the mean flow vorticity and the fluctuating vorticity [44, 45].

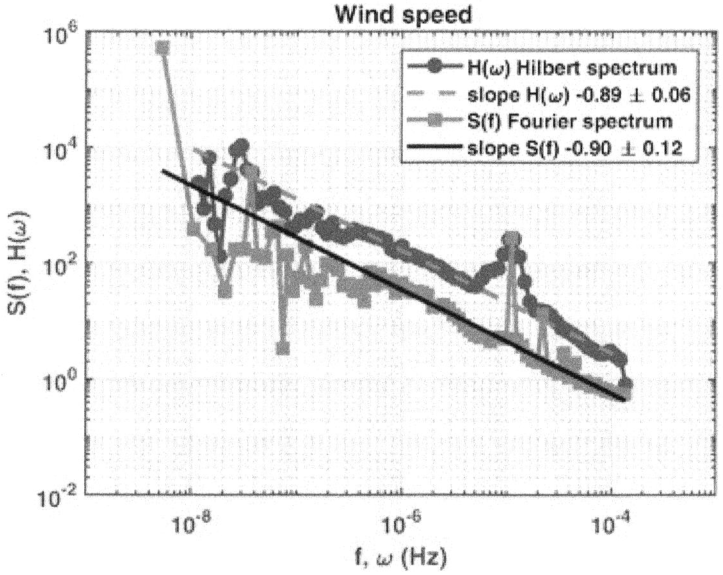

Figure 3. The spectrum of wind speed in the Hilbert space and the Fourier space. A power law behavior is observed in both spaces.

4.2 Time-frequency domain

The TFR in the Fourier and Hilbert spaces are, respectively, illustrated in Figures 4 and 5. Both figures show a color gradient from strong energy (in red) to weak energy (in blue). This highlights the energy activity related to PM10 concentrations during the study period. Such an approach gives the possibility of tracking the evolution of PM10 data spectral content in time, which is typically represented by variations of the amplitudes and frequencies of the components from which the signal is composed [46].

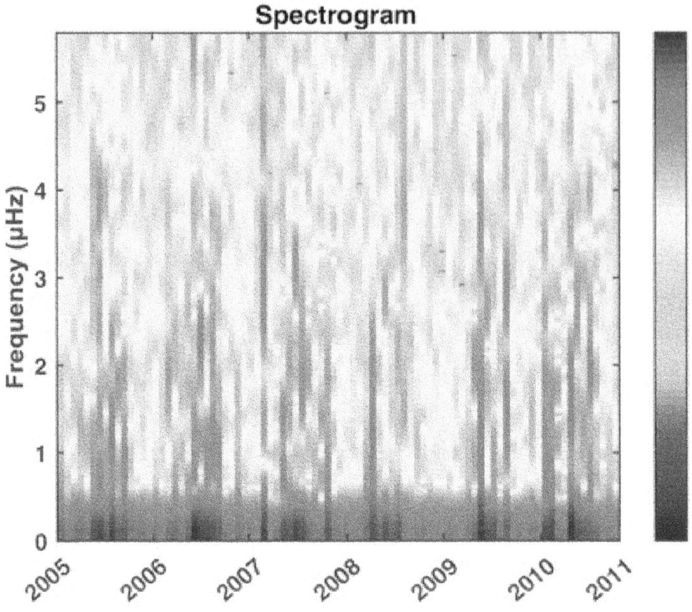

Figure 4. Spectrogram of PM10 time series with a color gradient from strong energy (in red) to weak energy (in blue).

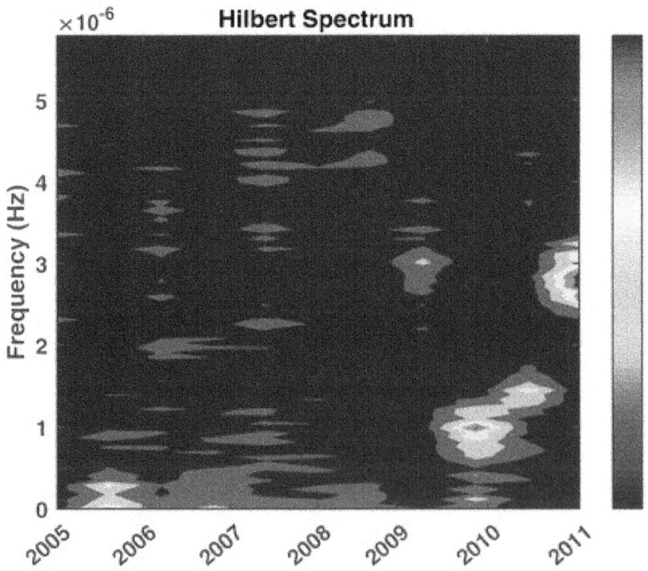

Figure 5. Hilbert spectrum of PM10 time series with a color gradient from strong energy (in red) to weak energy (in blue).

On Figure 4, strong energies are observed throughout the years with slight fluctuations on the frequency range $0 \leqslant f \leqslant 1\times10^{-6}$ Hz. For $f > 4\times10^{-6}$ Hz, strong energies are also noticed in the middle of each year and at the beginning of 2010 with more fluctuations. In Figure 5, energy distributions are more localized. On the frequency range $0 \leqslant f \leqslant 1\times10^{-6}$ Hz, we can observe the influence of small-scale event on energy behavior. As noticed, this energy may be weak or null. As an example, the impact of a general strike in early 2009 that paralyzed Guadeloupean archipelago at least 2 months is highlighted by zero energy due to the lack of PM10 sources, i.e., industrial activity and road traffic. For $f>1\times10^{-6}$ Hz, one can see more precisely energy variation related to dust events from mesoscale to large scale. Contrary to SPEC, HS clearly illustrates localized energy fluctuations due to small-scale event. In fact, the STFT makes an assumption that any signal as piecewise stationary and uses suitable window function to produce the short-time spectral characteristics of the signal. However, in reality, most of air pollution signals are usually nonstationary [9, 14, 47]. The Fourier transform-based technique treats the signal as a sum of predefined basis functions. If the analyzing signal is well matched with the bases, it performs better; otherwise the performance is degraded [10]. Here, the SPEC highlight energy fluctuations linked to PM10 coming from African dust between May and September (large-scale sources) [9] and from the eruption of Soufrière on Montserrat in February 2010 (mesoscale sources) [48]. However, the SPEC does not detect energy fluctuation related to anthropogenic pollution, i.e., local sources. This shows HS is a robust method in time-frequency domain. Indeed, based on the EMD method, this TFR is fully data adaptive, and the signal decomposition is performed without any predefined basis functions. These results confirm the superiority of HS over STFT in TFR.

5. CONCISE SUMMARY

This work examines the scaling and time-frequency characteristics of PM10 data using the Hilbert frame. The performances gained in the Hilbert space are being compared to those attained in the Fourier space. Using Hilbert spectral analysis (HSA), we observed a clear power law behavior in the frequency range of 2.09×10–7 to 4.57×10–5 Hz, which corresponds to time scales of 6.1 hours to 55.4 days. The estimated spectral exponent βh is 1.02 ± 0.10. Due to its strong localization in both physical and spectral domains, the HSA methodology incorporates the impact of sporadic dust events with significant variations in the amplitude-frequency space, which is not accounted for in the Fourier spectrum. Subsequently, the PM10 data is depicted using time-frequency representations utilizing the Hilbert spectrum and spectrogram. The results demonstrate that the HS-based TFR outperforms SPEC. The increased resolution in TFR provides improved variations of PM10 energy for frequencies below 1μ Hz. This is because it is not possible to enhance the TF resolution to the appropriate extent in SPEC. One significant advantage of HS is that it allows for a temporal resolution that is as accurate as the sampling period, while the frequency resolution can be chosen up to the Nyquist limit. Furthermore, unlike SPEC which results in a significant presence of cross-spectral energy components when using overlapping window functions, HS is completely adaptable to datasets because of its signal decomposition. These initial findings indicate a significant potential for conducting a comprehensive dynamic analysis of PM10 concentrations in the Caribbean region to accurately measure the source and level of pollution.

6. AT A GLANCE

Over the past two decades, numerous megacities have encountered air pollution, resulting in detrimental effects on human health. The presence of African dust in the Caribbean region significantly impacts air quality, leading to many diseases, including respiratory ailments. It is essential to enhance the comprehension of PM10 variations in order to develop methods and create tools for forecasting dust outbreaks. The initial phase involves characterizing the dynamic features of PM10 fluctuations, such as identifying potential scaling in the PM10 density power spectrum. The study examines the scale-invariant characteristics of PM10 daily time series over a period of 6 years using the theoretical Hilbert frame. Subsequently, the Hilbert spectrum is examined in the domain of time and frequency. The selection of the theoretical framework must be pertinent. The text includes a comparison of the results obtained in both the Hilbert and Fourier spaces.

REFERENCES

1. Xi W, Chen RJ, Chen BH, Kan HD. Application of statistical distribution of PM10 concentration in air quality management in 5 representative cities of China. Biomedical and Environmental Sciences. 2013;26(8):638-646
2. Plocoste T, Dorville JF, Monjoly S, Jacoby-Koaly S, André M. Assessment of nitrogen oxides and ground-level ozone behavior in a dense air quality station network: Case study in the Lesser Antilles arc. Journal of the Air & Waste Management Association. 2018;68(12):1278-1300
3. C, Gehrig R, Baltensperger U, Gysel M, Monn C, Vonmont H. Chemical characterisation of PM2.5, PM10 and coarse particles at urban, near-city and rural sites in Switzerland. Atmospheric Environment. 2005;39(4):637-651
4. M, Geravandi S, Rastegarimehr B, Valipour A, Ranjbarzadeh A, Yari AR, et al. An investigation of particulate matter and relevant cardiovascular risks in Abadan and Khorramshahr in 2014–2016. Toxin Reviews. 2018:1-8
5. Cairncross EK, John J, Zunckel M. A novel air pollution index based on the relative risk of daily mortality associated with short-term exposure to common air pollutants. Atmospheric Environment. 2007;41(38):8442-8454

6. P, Papaloukas C, Petrakis M, Karakitsios S. Assessment and prediction of short term hospital admissions: The case of Athens, Greece. Atmospheric Environment. 2008;42(30):7078-7086
7. S, Sicard P, Khaniabadi YO, De Marco A, Ghomeishi A, Goudarzi G, et al. A comparative study of hospital admissions for respiratory diseases during normal and dusty days in Iran. Environmental Science and Pollution Research. 2017;24(22):18152-18159
8. Khaniabadi YO, Fanelli R, De Marco A, Daryanoosh SM, Kloog I, Hopke PK, et al. Hospital admissions in Iran for cardiovascular and respiratory diseases attributed to the middle eastern dust storms. Environmental Science and Pollution Research. 2017;24(20):16860-16868
9. T, Calif R, Jacoby-Koaly S. Temporal multiscaling characteristics of particulate matter PM10 and ground-level ozone O_3 concentrations in Caribbean region. Atmospheric Environment. 2017;169:22-35
10. Molla MKI, Hirose K. Hilbert spectrum in time-frequency representation of audio signals considering disjoint orthogonality. Advances in Adaptive Data Analysis. 2010;2(03):313-336
11. Peng Z, Peter WT, Chu F. A comparison study of improved Hilbert-Huang transform and wavelet transform: Application to fault diagnosis for rolling bearing. Mechanical Systems and Signal Processing. 2005;19(5):974-988
12. Andrade AO, Kyberd P, Nasuto SJ. The application of the Hilbert spectrum to the analysis of electromyographic signals. Information Sciences. 2008;178(9):2176-2193
13. Plocoste T, Jacoby-Koaly S, Molinié J, Petit R. Evidence of the effect of an urban heat island on air quality near a landfill. Urban Climate. 2014;10:745-757
14. Plocoste T, Calif R, Jacoby-Koaly S. Multi-scale time dependent correlation between synchronous measurements of ground-level ozone and meteorological parameters in the Caribbean Basin. Atmospheric Environment. 2019;211:234-246
15. Euphrasie-Clotilde L, Molinié J, Feuillard T, Brute F. The relationship between coastal West African dust level and Caribbean island dust. WIT Transactions on Ecology and the Environment. 2017;211:121-127
16. Mandelbrot B. The Fractal Geometry of Nature. New York: Freeman and Company; 1982
17. Bracewell R. The Fourier Transform and Its Applications. 3rd ed. New York: McGraw-Hill Science; 1999

18. Sengupta R. Interpretation of magnetic anomalies of a two-dimensional fault by Fourier integral. Canadian Journal of Exploration Geophysicists. 1975;11:65-71

19. Huang NE, Shen Z, Long SR, Wu MC, Shih HH, Zheng Q, et al. The empirical mode decomposition and the Hilbert spectrum for nonlinear and non-stationary time series analysis. Proceedings of the Royal Society of London, Series A: Mathematical, Physical and Engineering Sciences. 1998;454(1971):903-995

20. Huang NE, Shen Z, Long SR. A new view of nonlinear water waves: The Hilbert spectrum. Annual Review of Fluid Mechanics. 1999;31(1):417-457

21. Flandrin P, Goncalves P. Empirical mode decompositions as data-driven wavelet-like expansions. International Journal of Wavelets, Multiresolution and Information Processing. 2004;2(04):477-496

22. Huang NE, Wu MLC, Long SR, Shen SS, Qu W, Gloersen P, et al. A confidence limit for the empirical mode decomposition and Hilbert spectral analysis. Proceedings of the Royal Society of London, Series A: Mathematical, Physical and Engineering Sciences. 2003;459(2037):2317-2345

23. Rilling G, Flandrin P. One or two frequencies? The empirical mode decomposition answers. IEEE Transactions on Signal Processing. 2008;56(1):85-95

24. Long SR, Huang N, Tung C, Wu M, Lin R, Mollo-Christensen E, et al. The Hilbert techniques: An alternate approach for non-steady time series analysis. GRSL Geoscience and Remote Sensing Letters. 1995;3:6-11

25. Cohen L. Time-Frequency Analysis. Vol. 778. New Jersey: Prentice Hall; 1995

26. Calif R, Schmitt FG, Huang Y. Multifractal description of wind power fluctuations using arbitrary order Hilbert spectral analysis. Physica A: Statistical Mechanics and its Applications. 2013;392(18):4106-4120

27. Calif R, Schmitt FG, Huang Y. The scaling properties of the turbulent wind using empirical mode decomposition and arbitrary order Hilbert spectral analysis. In: Wind Energy-Impact of Turbulence. Berlin: Springer; 2014. pp. 43-49

28. Schertzer D, Lovejoy S. Physical modeling and analysis of rain and clouds by anisotropic scaling multiplicative processes. Journal of Geophysical Research-Atmospheres. 1987;92:9693-9714

29. Marshak A, Davis A, Cahalan R, Wiscombe W. Bounded cascade models as nonstationary multifractals. Physical Review E. 1994;49:55-69

30. Windsor H, Toumi R. Scaling and persistence of UK pollution. Atmospheric Environment. 2001;35:4545-4556

31. Marr LC, Harley RA. Spectral analysis of weekday–weekend differences in ambient ozone, nitrogen oxide, and non-methane hydrocarbon time series in California. Atmospheric Environment. 2002;36:2327-2335

32. Messina AR, Vittal V. Nonlinear, non-stationary analysis of interarea oscillations via Hilbert spectral analysis. IEEE Transactions on Power Systems. 2006;21(3):1234-1241

33. Choi YS, Ho CH, Chen D, Noh YH, Song CK. Spectral analysis of weekly variation in PM10 mass concentration and meteorological conditions over China. Atmospheric Environment. 2008;42:655-666

34. Y, Schmitt FG. Time dependent intrinsic correlation analysis of temperature and dissolved oxygen time series using empirical mode decomposition. Journal of Marine Systems. 2014;130:90-100

35. Kingsbury BE, Morgan N, Greenberg S. Robust speech recognition using the modulation spectrogram. Speech Communication. 1998;25(1–3):117-132

36. Sussillo D, Kundaje A, Anastassiou D. Spectrogram analysis of genomes. EURASIP Journal on Advances in Signal Processing. 2004;2004(1):29-42

37. Fulop SA, Fitz K. Algorithms for computing the time-corrected instantaneous frequency (reassigned) spectrogram, with applications. The Journal of the Acoustical Society of America. 2006;119(1):360-371

38. Huang Y, Schmitt FG, Lu Z, Liu Y. An amplitude-frequency study of turbulent scaling intermittency using empirical mode decomposition and Hilbert spectral analysis. EPL (Europhysics Letters). 2008;84(4):40010

39. Ismail DKB, Lazure P, Puillat I. Advanced spectral analysis and cross correlation based on the empirical mode decomposition: Application to the environmental time series. IEEE Geoscience and Remote Sensing Letters. 2015;12(9):1968-1972

40. Li X, Li Z, Wang E, Feng J, Kong X, Chen L, et al. Analysis of natural mineral earthquake and blast based on Hilbert–Huang transform (HHT). Journal of Applied Geophysics. 2016;128:79-86

41. Kader B, Yaglom A. Spectra and correlation functions of surface layer atmospheric turbulence in unstable thermal stratification. In: Turbulence and Coherent Structures. Dordrecht: Springer; 1991. pp. 387-412

42. Huang Y, Schmitt FG, Hermand JP, Gagne Y, Lu Z, Liu Y, et al. Arbitrary-order Hilbert spectral analysis for time series possessing scaling statistics: Comparison study with detrended fluctuation analysis and wavelet leaders. Physical Review E. 2011;84(1):016208

43. Zhang C, Ni Z, Ni L. Multifractal detrended cross-correlation analysis between PM2.5 and meteorological factors. Physica A: Statistical Mechanics and its Applications. 2015;438:114-123

44. Tchen C. On the spectrum of energy in turbulent shear flow. Journal of Research of the National Bureau of Standards. 1953;50:51-62

45. Katul GG, Porporato A, Nikora V. Existence of k^{-1} power-law scaling in the equilibrium regions of wall-bounded turbulence explained by Heisenberg's eddy viscosity. Physical Review E. 2012;86(6):066311

46. Iatsenko D, McClintock PV, Stefanovska A. Linear and synchrosqueezed time–frequency representations revisited: Overview, standards of use, resolution, reconstruction, concentration, and algorithms. Digital Signal Processing. 2015;42:1-26

47. Varotsos C, Ondov J, Efstathiou M. Scaling properties of air pollution in Athens, Greece and Baltimore, Maryland. Atmospheric Environment. 2005;39(22):4041-4047

48. Prospero JM, Collard FX, Molinié J, Jeannot A. Characterizing the annual cycle of African dust transport to the Caribbean Basin and South America and its impact on the environment and air quality. Global Biogeochemical Cycles. 2014;28(7):757-773

08 OPTIMAL CONTROL OF EVOLUTIONARY DIFFERENTIAL INCLUSIONS USING POLYNOMIAL LINEAR DIFFERENTIAL OPERATORS

1. INTRODUCTION: AN OVERVIEW

This chapter focuses on a specific type of optimal control problem involving differential inclusions. In this case, the left-hand side of the evolution inclusion consists of polynomial linear differential operators with variable coefficients. The main challenge in these problems is to construct higher order adjoint inclusions of the Euler-Lagrange type and satisfy the transversality conditions. Therefore, the focus of the literature is solely on examining the qualitative characteristics of second-order differential inclusions. This can be observed in the works cited as references [1, 2, 3] and other relevant sources.

The work [1] provides the necessary and sufficient conditions that guarantee the existence of a solution to the second-order differential inclusion with a Cauchy initial value problem. Additionally, the concept of second-order interior tangent sets is introduced and analyzed in order to derive these criteria. The study [2] investigates the problem of three boundary conditions for second-order differential inclusions and second-order ordinary differential equations within the framework of Banach spaces. The results are derived in several novel contexts of Sobolev-type spaces that incorporate Bochner and Pettis integrals. The study [3] demonstrates the existence of feasible solutions to the Cauchy problem $x''\in F(x,x'),x(0)=x_0,x'(0)=y_0$, where F is a set-valued map formed on a locally compact set $M\subset\mathbb{R}^{2n}$, which is a subset of the Frechet subdifferential of a φ-convex function of order two.

The study of qualitative properties and optimization of first-order discrete and continuous time processes with lumped and distributed parameters has experienced significant growth in recent decades. This expansion is evident in the extensive literature, as evidenced by the numerous references cited in [4, 5, 6, 7, 8, 9, 10, 11, 12, 13].

Mahmudov pioneered the optimization of higher order differential inclusions in his works [14, 15, 16, 17, 18, 19, 20, 21]. Since then, this subject has garnered significant attention from other authors (see to [22] and its references). The work [14] examines a novel category of optimal control theory problems that involve Sturm-Liouville type differential inclusions. These inclusions incorporate second-order linear self-adjoint differential operators. The discretization method is employed to ensure a smooth transition to the continuous problem, allowing for the investigation of both discrete and discrete-approximate inclusions. The derivation includes necessary and sufficient criteria that encompass both the Euler-Lagrange and Hamiltonian type inclusions, as well as "transversality" conditions. The work [15] focuses on optimizing the Bolza issue by considering third-order differential inclusions and arbitrarily higher order discrete inclusions. The study [16] focuses on the Bolza problem in optimal control theory, which involves second-order convex differential inclusions with second-order state variable inequality requirements. The suggested discretization method examines situations involving discrete-approximate inclusions and inequalities. The necessary and sufficient criteria for optimality, including the specific "transversality" criterion, are proven in the form of Euler-Lagrange inclusions.

The work [17] examines the essential and complete requirements for achieving the best possible outcome in second-order polyhedral optimization, which is characterized by polyhedral discrete and differential inclusions. The work [18] focuses on the examination of optimal control theory including higher order differential inclusions and a time interval that varies. In a broader context of issues and limitations, the primary objective is to determine adequate criteria for optimality in higher order differential inclusions. Therefore, by employing Euler-Lagrange and Hamiltonian inclusions, along with transversal conditions on the "initial" sets, the necessary conditions can be expressed. The work [21] examines a novel category of optimal control theory problems involving state restrictions and second-order delay discrete and delay differential inclusions. The optimality conditions for delay differential functional inclusions (DFIs), including the specific transversality conditions, are proven by utilizing discrete approximations in the form of Euler-Lagrange and Hamiltonian type inclusions, under the assumption of "regularity".

To establish the Euler-Lagrangess and Hamiltonian inclusions and the transversality conditions, we use the construction of a suitable rewriting of the primal polynomial linear differential operator and the rearrangement of its integration. The case of variable coefficients of polynomial linear differential operators turns out to be more complicated, unless transversality assumptions at the endpoints of the considered time interval are applied. It should be noted that the main proof can be easily generalized to the nonconvex case. Then, using the new approach given in Section 4 of this chapter, we construct the Weierstrass-Pontryagin maximum condition [24] for the linear optimal control problem. Consequently, in the particular case, the maximum principle follows from the Euler-Lagrange inclusion.

In Section 5 the optimality conditions are given for convex problem with second-order differential inclusions and endpoint constraints. By using second-order suitable Euler-Lagrange type adjoint inclusions and transversality conditions, Theorem 5.1 is proved.

The main results in this section can be extended to the case of Hilbert spaces ℓ_2, L^n_2. We remind that a Hilbert space H is a real or complex inner product space that is also a complete metric space with respect to the distance function induced by the inner product [2]. By definition, every Hilbert space is also a Banach space. Furthermore, in every Hilbert space, the following parallelogram identity $\|x+y\|^2+\|x-y\|^2=2\left(\|x\|^2+\|y\|^2\right)$ holds. Conversely, every Banach space in which the parallelogram identity holds is a Hilbert space. Remember that ℓ_2 is a space of numerical sequences, such that if $x = \{x_i\}$, then $\sum_{i=1}^{\infty} x_i^2 < \infty$.. In fact ℓ_2 is an infinity dimensional coordinate-wise Hilbert space with the corresponding inner product $\langle x, y\rangle = \sum_{i=1}^{\infty} x_i y_i$.. Endowing a relevant norm, we have a Banach space. Obviously, optimization of problem with PLDOs can be reduced to problem with geometric constraints in such finite-dimensional Hilbert space. As is known with all the pairs of elements of this space, a certain finite number is associated, i.e., inner product, existence of which is guaranteed by applying the familiar Cauchy Schwarz-Bunyakovskii [25] inequality. We remark that in our case for $x=\{x_0,x_1,x_2,\ldots\}\in \ell_2$ and $x^*=\{x_0^*, x_1^*, x_2^*, \ldots\}\in \ell_2^*$, the inner product $\langle x, x^*\rangle = \sum_{i=1}^{\infty} x_i x_i^*$ is finite numbers since this series is convergent. Besides it is known [25] that ℓ_2 is a self-adjoint space, i.e., $\ell_p=\ell^*_q$, and $1/p+1/q=1$, and so $\ell_2=\ell^*_2$ for p=2. Thus a dual cone constructed can be defined. The set of square integrable functions $L^n_2([0,1])$ is a Hilbert space with inner product $\langle x(t), y(t)\rangle = \int_0^1 x(t)y(t)dt$.

2. PRELIMINARIES AND PROBLEM STATEMENTS

The basic concepts given in this section can be found in the book [23]; let $\mathbb{R}^n$ be a n-dimensional Euclidean space, $\langle x,v\rangle$ be an inner product of elements $x,v\in \mathbb{R}^n$, and (x,v) be a pair of x,v. Let $F: \mathbb{R}^n \rightrightarrows \mathbb{R}^n$ be a set-valued mapping from $\mathbb{R}^n$ into the set of subsets of $\mathbb{R}^n$. Therefore F is a convex set-valued mapping, if its graph gph $F=\{(x,v):v\in F(x)\}$ is a convex subset of $\mathbb{R}^{2n}$. A set-valued mapping F is called closed if its gph$F$ is a closed subset in $\mathbb{R}^{2n}$. The domain of a set-valued mapping F is denoted by dom F and is defined as dom $F=\{x:F(x)\neq\emptyset\}$. A set-valued mapping F is convex-valued if $F(x)$ is a convex set for each $x\in$ dom$F$.

The Hamiltonian function and argmaximum set corresponding to a set-valued mapping F are defined by the relations correspondingly:

$$H_F(x,v^*) = \sup_v\{\langle v,v^*\rangle : v\in F(x)\},\ \ v^*\in\mathbb{R}^n,$$

$$F_A(x,v^*) = \{v\in F(x) : \langle v,v^*\rangle = H_F(x,v^*)\}$$

We set $H_F(x,v^*)=-\infty$ if $F(x)=\emptyset$. The interior and relative interior of a set $M\subset\mathbb{R}^{2n}$ are denoted by int$M$ and ri$M$, respectively.

A convex cone $K_M(z_0)$, $z_0\in M$ is a cone of tangent directions if from $\bar{z} = (\bar{x},\bar{v})\in K_M(z_0)$ it follows that $\bar{z}$ is a tangent vector to the set M at a point $z_0\in M$, i.e., there exists such function q: $\mathbb{R}^1\rightarrow\mathbb{R}^{2n}$ that $z_0+\alpha\bar{z}+q(\alpha)\in M$ for sufficiently small $a > 0$ and $a^{-1}q(a)\rightarrow 0$, as $a\downarrow 0$.

For a set-valued mapping F, the set-valued mapping F^*: $\mathbb{R}^n\rightrightarrows\mathbb{R}^n$ is defined by

$$F^*(v^*;(x,v)):=\{x^* : (x^*,-v^*)\in K^*_{gph\,F}(x,v)\}$$

$$K_{gphF}(x,v) = \text{cone}\,[\,gph\,F-(x,v)],\forall(x^1,v^1)\in gph\,F.$$

It is called the LAM to F at a point $(x,v)\in gph\ F$, where $K^* = \{z^* : \langle\bar{z},z^*\rangle\geq 0,\ \forall\bar{z}\in K\}$ denotes the dual cone to the cone K, as usual. Below by using the Hamiltonian function, associated to a set-valued mapping F, we will define another LAM. Thus, the LAM to "nonconvex" mapping F is defined as follows:

$$F^*(v^*;(x,v)):=\{x^* : H_F(x^1,v^*)-H_F(x,v^*)\leq\langle x^*,x^1-x\rangle,\forall x^1\in\mathbb{R}^n\},$$
$$(x,v)\in gph\,F, v\in F_A(x,v^*).$$

Clearly, for the convex mapping F, the Hamiltonian function $H_F(\cdot,\cdot,v^*)$ is concave, and the latter definition of LAM coincides with the previous definition of LAM ([23], p. 62). Note that prior to the LAM, the notion of coderivative has been introduced for set-valued mappings in terms of the basic normal cone to their graphs by Mordukhovich [26] and for the smooth convex maps, the two notions are equivalent.

The aim of Section 3 is to obtain the Euler-Lagrange type adjoint inclusion and sufficient optimality conditions for a problem with polynomial linear differential operators:

$$\text{Minimize } \varphi\left(x(1),x'(1),\ldots,x^{(s-1)}(1)\right),\qquad(1)$$

$$(P_V)\ Lx(t)\in F(x(t),t),\ \text{a.e. } t\in[0,1],\qquad(2)$$

$$x(0)\in Q_0,\ x'(0)\in Q_1 x''(0)\in Q_2,\ \ldots,x^{(s-1)}(0)\in Q_{s-1}\qquad(3)$$

where $Lx=\sum_{k=1}^{s}p_k(t)D^k x$ is a PLDO of degree s with variable coefficients pk:[0,1]→$\mathbb{R}^1$ and D^k,k=1,...,s is the operator of kth-order derivatives. In what follows for each k, a scalar function p_k is kth-order continuously differentiable function, $p_s(t)\neq 0$ on [0,1]01 identically, F(·,t): $\mathbb{R}^n\rightrightarrows\mathbb{R}^n$ is time-dependent set-valued mapping, φ:$(\mathbb{R}^n)^s\rightarrow\mathbb{R}^1$ is continuous function, $Q_j\subseteq\mathbb{R}^n$, j=0,1,...,s-1 are nonempty subsets of

$\mathbb{R}^n$, and $s(s\geq 2)$ is an arbitrary fixed natural number. It is required to find an arc $x(t)$ of the problem Eqs. (1)–(3) for the sth-order differential inclusions satisfying Eq. (2) almost everywhere (a.e.) on a considered time interval and minimizing the functional $\varphi(x(1),\ldots,x^{(s-1)}(1))$ An arc $x(\cdot)$·is absolutely continuous and s-1 order differentiable function, where $x^{(s)}(\cdot) \equiv \frac{d^s x(\cdot)}{dt^s} \in L_1^n([0,1])$. Obviously, such class of functions is a Banach space, endowed with the different equivalent norms.

Remark 2.1. Notice that to get sufficient condition of optimality for the Mayer problem (P_V) described by ordinary evolution differential inclusions with PLDOs and with initial point constraints, using the discretized method, we consider the following sth-order discrete-approximate problem instead of the problem (P_V):

$$\text{minimize} \quad \varphi(x(1-(s-1)h), \Delta x(1-(s-1)h), \ldots, \Delta^{s-1}x(1-(s-1)h)),$$

$$\sum_{k=1}^{s} p_k(t)\Delta^k x(t) \in F(x(t),t), t = 0, h, 2h, \ldots, 1-sh;$$

$$\Delta^k x(0) \in Q_k, k = 0, \ldots, s-1.$$

Here kth-order difference operator is defined as follows:

$$\Delta^k x(t) = \frac{1}{h^k}\sum_{s=0}^{k}(-1)^s C_k^s x(t+(k-s)h), \quad C_k^s = \frac{k!}{s!(k-s)!}, \quad t = 0, h, \ldots, 1-h.$$

Thus by using the method of approximation [23, 26, 27], we can establish necessary and sufficient conditions for the rather complicated sth-order discrete-approximate problem. Then by passing to the limit in necessary and sufficient conditions of this problem as $h\to 0$, we can construct the optimality condition for the Mayer problem (P_V) described by higher order differential inclusions with PLDOs and with initial point constraints. But in this chapter to avoid long calculations, derivations of these conditions are omitted.

3. OPTIMIZATION OF EVOLUTION DIFFERENTIAL INCLUSIONS WITH PLDOS

In the present section, we study sufficient optimality conditions for the problem (P_V). Before all, we formulate the so-called sth-order Euler-Lagrange type differential inclusion and the transversality conditions:

$$\text{(i)} \; L^* x^*(t) \in F^*(x^*(t); (\tilde{x}(t), L\tilde{x}(t)), t), \text{ a.e. } t \in [0,1],$$

where $L^* x^*(t) = \sum_{k=1}^{s}(-1)^k D^k[p_k(t)x^*(t)]$ is the adjoint PLDO of the primal operator L.

(ii) $\sum_{k=0}^{s-1}(-1)^{s-k}D^{s-k-1}[p_{s-k}(0)x^*(0)] \in K^*_{Q_0}(\tilde{x}(0))$;

$\sum_{k=0}^{s-2}(-1)^{s-k-1}D^{s-k-2}[p_{s-k}(0)x^*(0)] \in K^*_{Q_1}(\tilde{x}'(0))$;

$D[p_s(0)x^*(0)] - p_{s-1}(0)x^*(0) \in K^*_{Q_{s-2}}\left(\tilde{x}^{(s-2)}(0)\right)$;

$-p_s(0)x^*(0) \in K^*_{Q_{s-1}}\left(\tilde{x}^{(s-1)}(0)\right)$.

(iii) $\left(\sum_{k=0}^{s-1}(-1)^{s-k}D^{s-k-1}[p_{s-k}(1)x^*(1)], \sum_{k=0}^{s-2}(-1)^{s-k-1}D^{s-k-2}[p_{s-k}(1)x^*(1)], \ldots,\right.$
$\left. D[p_s(1)x^*(1)] - p_{s-1}(1)x^*(1), p_s(1)x^*(1)\right) \in \partial\varphi\left(\tilde{x}(1), \tilde{x}'(1), \ldots, \tilde{x}^{(s-1)}(1)\right)$.

Later on we assume that $x^*(t), t\in[0,1]$ is absolutely a continuous function with the higher order derivatives until s-1 and $x^{*(s)}(\cdot) \in L_1^n([0,1])$. The following condition ensures that the LAM F^* is nonempty:

(iv) $L\tilde{x}(t) \in F_A(\tilde{x}(t), x^*(t), t)$, a.e. $t \in [0,1]$ or, equivalently,

$$\langle L\tilde{x}(t), x^*(t)\rangle = H_F(\tilde{x}(t), x^*(t)),\ L\tilde{x}(t) \in F(\tilde{x}(t), t).$$

The following are sufficient optimality conditions for evolution differential inclusions with PLDOs.

Theorem 3.1. Let φ be a continuous and convex function and $F(\cdot,t)$ a convex set-valued mapping. Moreover, let

$$Q_j, j = 0, \ldots, s-1$$

be convex sets. Then for optimality of the trajectory $\tilde{x}(\cdot)$ in the problem (P_V) with evolution differential inclusions and PLDOs, it is sufficient that there exists an absolutely continuous function $x^*(\cdot)$ with the higher order derivatives until s-1, satisfying a.e. the Euler-Lagrange type differential inclusion with PLDOs Eqs. (i) and (iv) and transversality conditions Eqs. (ii) and (iii) at the endpoints t=0 and t=1.

Proof. Using Theorem 2.1 ([23], p. 62), the definition of the Hamiltonian function and condition Eq. (i), we obtain

$$H_F(x(t), x^*(t), t) - H_F(\tilde{x}(t), x^*(t), t) \le \langle L^*x^*(t), x(t) - \tilde{x}(t)\rangle$$

which can be rewritten as follows

$$H_F(x(t), x^*(t), t) - H_F(\tilde{x}(t), x^*(t), t)$$

$$\le \left\langle \sum_{k=1}^{s}(-1)^k \left(p_k(t)x^*(t)\right)^{(k)}, x(t) - \tilde{x}(t)\right\rangle. \tag{4}$$

Further using the definition of the Hamiltonian function, Eq. (4) can be converted to the inequality

$$0 \geq \langle Lx(t) - L\tilde{x}(t), x^*(t)\rangle - \langle L^*x^*(t), x(t) - \tilde{x}(t)\rangle$$

or

$$0 \geq \left\langle \sum_{k=1}^{s} p_k(t)(x(t) - \tilde{x}(t))^{(k)}, x^*(t) \right\rangle - \left\langle \sum_{k=1}^{s} (-1)^k \left(p_k(t)x^*(t)\right)^{(k)}, x(t) - \tilde{x}(t) \right\rangle. \tag{5}$$

Integrating Eq. (5) over the interval [0,1]01, we have

$$0 \geq \int_0^1 \left[\left\langle \sum_{k=1}^{s} p_k(t)(x(t) - \tilde{x}(t))^{(k)}, x^*(t) \right\rangle \right.$$

$$\left. - \left\langle \sum_{k=1}^{s} (-1)^k \left(p_k(t)x^*(t)\right)^{(k)}, x(t) - \tilde{x}(t) \right\rangle \right] dt \tag{6}$$

Let us denote

$$B = \sum_{k=1}^{s} \left\langle (x(t) - \tilde{x}(t))^{(k)}, p_k(t)x^*(t) \right\rangle - \sum_{k=1}^{s} \left\langle (-1)^k \left(p_k(t)x^*(t)\right)^{(k)}, x(t) - \tilde{x}(t) \right\rangle$$

In what follows our approach lies in reducing B in a relationship consisting of s sums from $k(k=1,\ldots,s)$ to s of suitable derivatives of scalar products; thus, after some transformations we can deduce an important representation for a first term of B as follows:

$$\begin{aligned}
\sum_{k=1}^{s} \left\langle (x(t) - \tilde{x}(t))^{(k)}, p_k(t)x^*(t) \right\rangle &= \sum_{k=1}^{s} \left[\frac{d}{dt} \left\langle x^{(k-1)}(t) - \tilde{x}^{(k-1)}(t), p_k(t)x^*(t) \right\rangle \right] \\
&- \sum_{k=2}^{s} \left[\frac{d}{dt} \left\langle x^{(k-2)}(t) - \tilde{x}^{(k-2)}(t), \left(p_k(t)x^*(t)\right)' \right\rangle \right] \\
&+ \sum_{k=3}^{s} \left[\frac{d}{dt} \left\langle x^{(k-3)}(t) - \tilde{x}^{(k-3)}(t), \left(p_k(t)x^*(t)\right)'' \right\rangle \right] \\
- \cdots + &\sum_{k=s-2}^{s} \left[\frac{d}{dt} \left\langle x^{(k-s+2)}(t) - \tilde{x}^{(k-s+2)}(t), (-1)^{m-3} \left(p_k(t)x^*(t)\right)^{(s-3)} \right\rangle \right] \\
&+ \sum_{k=s-1}^{s} \left[\frac{d}{dt} \left\langle x^{(k-s+1)}(t) - \tilde{x}^{(k-s+1)}(t), (-1)^{s-2} \left(p_k(t)x^*(t)\right)^{(s-2)} \right\rangle \right] \\
&+ \frac{d}{dt} \left\langle x(t) - \tilde{x}(t), (-1)^{m-1} \left(p_s(t)x^*(t)\right)^{(s-1)} \right\rangle \\
&+ \sum_{k=1}^{s} \left[\left\langle x(t) - \tilde{x}(t), (-1)^k \left(p_k(t)x^*(t)\right)^{(k)} \right\rangle \right].
\end{aligned} \tag{7}$$

Then in view of Eq. (7) in the definition of B, we have an efficient formula:

$$B=\sum_{k=1}^{s}\left[\frac{d}{dt}\left\langle x^{(k-1)}(t)-\tilde{x}^{(k-1)}(t),p_k(t)x^*(t)\right\rangle\right]$$

$$-\sum_{k=2}^{s}\left[\frac{d}{dt}\left\langle x^{(k-2)}(t)-\tilde{x}^{(k-2)}(t),\left(p_k(t)x^*(t)\right)'\right\rangle\right] \tag{8}$$

$$+\sum_{k=3}^{s}\left[\frac{d}{dt}\left\langle x^{(k-3)}(t)-\tilde{x}^{(k-3)}(t),\left(p_k(t)x^*(t)\right)''\right\rangle\right]$$

$$-\cdot\cdots+\sum_{k=s-2}^{s}\left[\frac{d}{dt}\left\langle x^{(k-s+2)}(t)-\tilde{x}^{(k-s+2)}(t),(-1)^{s-3}\left(p_k(t)x^*(t)\right)^{(s-3)}\right\rangle\right]$$

$$+\sum_{k=s-1}^{s}\left[\frac{d}{dt}\left\langle x^{(k-s+1)}(t)-\tilde{x}^{(k-s+1)}(t),(-1)^{s-2}\left(p_k(t)x^*(t)\right)^{(s-2)}\right\rangle\right]$$

$$+\frac{d}{dt}\left\langle x(t)-\tilde{x}(t),(-1)^{s-1}\left(p_s(t)x^*(t)\right)^{(s-1)}\right\rangle.$$

Then taking into account the structure of B in Eq. (8), we can compute the integral on the right-hand side of Eq. (6) as follows:

$$\int_0^1 Bdt=\sum_{k=1}^{s}\left[\int_0^1 d\left\langle x^{(k-1)}(t)-\tilde{x}^{(k-1)}(t),p_k(t)x^*(t)\right\rangle\right]$$

$$-\sum_{k=2}^{s}\left[\int_0^1 d\left\langle x^{(k-2)}(t)-\tilde{x}^{(k-2)}(t),\left(p_k(t)x^*(t)\right)'\right\rangle\right]$$

$$+\sum_{k=3}^{s}\left[\int_0^1 d\left\langle x^{(k-3)}(t)-\tilde{x}^{(k-3)}(t),\left(p_k(t)x^*(t)\right)''\right\rangle\right]$$

$$-\cdot\cdots+\sum_{k=s-2}^{s}\left[\int_0^1 d\left\langle x^{(k-s+2)}(t)-\tilde{x}^{(k-s+2)}(t),(-1)^{s-3}\left(p_k(t)x^*(t)\right)^{(s-3)}\right\rangle\right]$$

$$+\sum_{k=s-1}^{s}\left[\int_0^1 d\left\langle x^{(k-s+1)}(t)-\tilde{x}^{(k-s+1)}(t),(-1)^{s-2}\left(p_k(t)x^*(t)\right)^{(s-2)}\right\rangle\right]$$

$$+\int_0^1 d\left\langle x(t)-\tilde{x}(t),(-1)^{s-1}\left(p_s(t)x^*(t)\right)^{(s-1)}\right\rangle.$$

Thus, integrating B, we can obtain

$$\int_0^1 Bdt=\sum_{k=1}^{s}\left[\left\langle x^{(k-1)}(1)-\tilde{x}^{(k-1)}(1),p_k(1)x^*(1)\right\rangle\right.$$

$$\left.-\left\langle x^{(k-1)}(0)-\tilde{x}^{(k-1)}(0),p_k(0)x^*(0)\right\rangle\right]$$

$$-\sum_{k=2}^{s}\left[\left\langle x^{(k-2)}(1)-\tilde{x}^{(k-2)}(1),\left(p_k(1)x^*(1)\right)'\right\rangle-\left\langle x^{(k-2)}(0)-\tilde{x}^{(k-2)}(0),\left(p_k(0)x^*(0)\right)'\right\rangle\right]$$

$$+\sum_{k=3}^{s}\left[\left\langle x^{(k-3)}(1)-\tilde{x}^{(k-3)}(1),\left(p_k(1)x^*(1)\right)''\right\rangle\right.$$

$$\left.-\left\langle x^{(k-3)}(0)-\tilde{x}^{(k-3)}(0),\left(p_k(0)x^*(0)\right)''\right\rangle\right]$$

$$-\cdots+\sum_{k=s-2}^{s}\left[\left\langle x^{(k-s+2)}(1)-\tilde{x}^{(k-s+2)}(1),(-1)^{s-3}\left(p_k(1)x^*(1)\right)^{(s-3)}\right\rangle\right.$$

$$\left.-\left\langle x^{(k-s+2)}(0)-\tilde{x}^{(k-s+2)}(0),(-1)^{s-3}\left(p_k(0)x^*(0)\right)^{(s-3)}\right\rangle\right]$$

$$+\sum_{k=s-1}^{s}\left[\left\langle x^{(k-s+1)}(1)-\tilde{x}^{(k-s+1)}(1),(-1)^{s-2}\left(p_k(1)x^*(1)\right)^{(s-2)}\right\rangle\right.$$

$$\left.-\left\langle x^{(k-s+1)}(0)-\tilde{x}^{(k-s+1)}(0),(-1)^{s-2}\left(p_k(0)x^*(0)\right)^{(s-2)}\right\rangle\right]$$

$$+\left\langle x(1)-\tilde{x}(1),(-1)^{s-1}\left(p_s(1)x^*(1)\right)^{(s-1)}\right\rangle$$

$$-\left\langle x(0)-\tilde{x}(0),(-1)^{s-1}\left(p_s(0)x^*(0)\right)^{(s-1)}\right\rangle.$$

Here by suitable rearrangement and necessary simplification, we have

$$\int_0^1 Bdt=\sum_{k=1}^{s}\left[\left\langle x^{(k-1)}(1)-\tilde{x}^{(k-1)}(1),p_k(1)x^*(1)\right\rangle\right]$$

$$-\sum_{k=2}^{s}\left[\left\langle x^{(k-2)}(1)-\tilde{x}^{(k-2)}(1),\left(p_k(1)x^*(1)\right)'\right\rangle\right]$$

$$+\sum_{k=3}^{s}\left[\left\langle x^{(k-3)}(1)-\tilde{x}^{(k-3)}(1),\left(p_k(1)x^*(1)\right)''\right\rangle\right]$$

$$
\begin{aligned}
&-\cdots+\sum_{k=s-2}^{s}\left[\left\langle x^{(k-s+2)}(1)-\tilde{x}^{(k-s+2)}(1),(-1)^{s-3}\left(p_k(1)x^*(1)\right)^{(s-3)}\right\rangle\right]\\
&+\sum_{k=s-1}^{s}\left[\left\langle x^{(k-s+1)}(1)-\tilde{x}^{(k-s+1)}(1),(-1)^{s-2}\left(p_k(1)x^*(1)\right)^{(s-2)}\right\rangle\right]\\
&+\left\langle x(1)-\tilde{x}(1),(-1)^{s-1}\left(p_s(1)x^*(1)\right)^{(s-1)}\right\rangle\\
&+\left\langle x(0)-\tilde{x}(0),\sum_{k=0}^{s-1}(-1)^{s-k}D^{s-k-1}\left[p_{s-k}(0)x^*(0)\right]\right\rangle\\
&+\left\langle x'(0)-\tilde{x}'(0),\sum_{k=0}^{s-2}(-1)^{s-k-1}D^{s-k-2}\left[p_{s-k}(0)x^*(0)\right]\right\rangle\\
&+\left\langle x''(0)-\tilde{x}''(0),\sum_{k=0}^{s-3}(-1)^{s-k-2}D^{s-k-3}\left[p_{s-k}(0)x^*(0)\right]\right\rangle\\
&+\cdot\cdots-\left\langle x^{(s-2)}(0)-\tilde{x}^{(s-2)}(0),-D\left[p_s(0)x^*(0)\right]+p_{s-1}(0)x^*(0)\right\rangle\\
&-\left\langle x^{(s-1)}(0)-\tilde{x}^{(s-1)}(0),p_s(0)x^*(0)\right\rangle.
\end{aligned}
\tag{9}
$$

In order to make use of the transversality condition Eq. (ii), we rewrite it in a more relevant form:

$$
\begin{aligned}
&\left\langle x(0)-\tilde{x}(0),\sum_{k=0}^{s-1}(-1)^{s-k}D^{s-k-1}\left[p_{s-k}(0)x^*(0)\right]\right\rangle\\
&+\left\langle x'(0)-\tilde{x}'(0),\sum_{k=0}^{s-2}(-1)^{s-k-1}D^{s-k-2}\left[p_{s-k}(0)x^*(0)\right]\right\rangle\\
&+\left\langle x''(0)-\tilde{x}''(0),\sum_{k=0}^{s-3}(-1)^{s-k-2}D^{s-k-3}\left[p_{s-k}(0)x^*(0)\right]\right\rangle\\
&+\cdot\cdots-\left\langle x^{(s-2)}(0)-\tilde{x}^{(s-2)}(0),-D\left[p_s(0)x^*(0)\right]+p_{s-1}(0)x^*(0)\right\rangle\\
&-\left\langle x^{(s-1)}(0)-\tilde{x}^{(s-1)}(0),p_s(0)x^*(0)\right\rangle\geq 0;\forall x^{(k)}(0)\in K_{Q_k}\left(\tilde{x}^{(k)}(0)\right),
\end{aligned}
$$

$k=0,\ldots,s-1$. Thus, from Eqs. (6) and (9), we have

$$
\begin{aligned}
0 \geq & \sum_{k=1}^{s}\left[\left\langle x^{(k-1)}(1)-\tilde{x}^{(k-1)}(1), p_k(1)x^*(1)\right\rangle\right] \\
& -\sum_{k=2}^{s}\left\langle x^{(k-2)}(1)-\tilde{x}^{(k-2)}(1), \frac{d(p_k(1)x^*(1))}{dt}\right\rangle \\
& +\sum_{k=3}^{s}\left[\left\langle x^{(k-3)}(1)-\tilde{x}^{(k-3)}(1), \frac{d^2(p_k(1)x^*(1))}{dt^2}\right\rangle\right]-\cdots \\
& +\sum_{k=s-2}^{s}\left[\left\langle x^{(k-s+2)}(1)-\tilde{x}^{(k-s+2)}(1),(-1)^{s-3}\frac{d^{s-3}(p_k(1)x^*(1))}{dt^{s-3}}\right\rangle\right] \\
& +\sum_{k=s-1}^{s}\left[\left\langle x^{(k-s+1)}(1)-\tilde{x}^{(k-s+1)}(1),(-1)^{s-2}\frac{d^{s-2}(p_k(1)x^*(1))}{dt^{s-2}}\right\rangle\right] \\
& +\left\langle x(1)-\tilde{x}(1),(-1)^{s-1}\frac{d^{s-1}(p_s(1)x^*(1))}{dt^{s-1}}\right\rangle.
\end{aligned}
$$

Using the derivative operator D, it is not hard to see that the relation described above can be expressed in a more compact form:

$$
\begin{aligned}
0 \geq & \left\langle \sum_{k=0}^{s-1}(-1)^{s-k-1}D^{s-k-1}\left[p_{s-k}(1)x^*(1)\right], x(1)-\tilde{x}(1)\right\rangle \\
& +\left\langle \sum_{k=0}^{s-2}(-1)^{s-k-2}D^{s-k-2}\left[p_{s-k}(1)x^*(1)\right], x'(1)-\tilde{x}'(1)\right\rangle+\cdots \qquad (10)\\
& -\left\langle D\left[p_s(1)x^*(1)\right]-p_{s-1}(1)x^*(1), x^{(s-2)}(1)-\tilde{x}^{(s-2)}(1)\right\rangle \\
& +\left\langle p_s(1)x^*(1), x^{(s-1)}(1)-\tilde{x}^{(s-1)}(1)\right\rangle.
\end{aligned}
$$

Furthermore, applying the definition of the transversality condition Eq. (iii) for all feasible arc $x(\cdot)$ we have

$$
\begin{aligned}
& \varphi\left(x(1),x'(1),\ldots,x^{(s-1)}(1)\right)-\varphi\left(\tilde{x}(1),\tilde{x}'(1),\ldots,\tilde{x}^{(s-1)}(1)\right) \\
& \geq \left\langle \sum_{k=0}^{s-1}(-1)^{s-k}D^{s-k-1}\left[p_{s-k}(1)x^*(1)\right], x(1)-\tilde{x}(1)\right\rangle \\
& +\left\langle \sum_{k=0}^{s-2}(-1)^{s-k-1}D^{s-k-2}\left[p_{s-k}(1)x^*(1)\right], x'(1)-\tilde{x}'(1)\right\rangle+\cdots
\end{aligned}
$$

$$+\left\langle D\left[p_s(1)x^*(1)\right]-p_{s-1}(1)x^*(1),x^{(s-2)}(1)-\tilde{x}^{(s-2)}(1)\right\rangle$$

$$-\left\langle p_s(1)x^*(1),x^{(s-1)}(1)-\tilde{x}^{(s-1)}(1)\right\rangle \tag{11}$$

Then from the last two inequalities Eqs. (10) and (11) for all feasible arc $x(\cdot)$ we have immediately $\varphi(x(1),x'(1),\ldots,x^{(s-1)}(1))\geq\varphi\left(\tilde{x}(1),\tilde{x}'(1),\ldots,\tilde{x}^{(s-1)}(1)\right)$,that is, $x(\cdot)$ is an optimal trajectory.

Remark 3.1. It can be noted that in the particular case, if $p_2(t)=p(t)$, $p_1(t)=p'(t)$, where $p(\cdot)$: $[0,1]\to(0,\infty)$, the second-order linear differential operator $Lx=p^2(t)x''+p^1(t)x'$ is a well-known self-adjoint Sturm-Liouville operator $Lx\equiv(px')'$.

Corollary 3.1. Let $F(\cdot,t)$ be a closed set-valued mapping. Then under the assumptions of Theorem 3.1, the conditions Eqs. (i) and (iii) can be rewritten in terms of Hamiltonian function as follows:

$$L^*x^*(t)\in\partial_x H_F(\tilde{x}(t),x^*(t),t);\ .L\tilde{x}(t)\in\partial_{v^*}H(\tilde{x}(t),x^*(t),t),\ \text{a.e.}t\in[0,1]$$

Proof. Indeed, by Theorem 2.1 ([23], p. 62) and Lemma 5.1 [14], we can write.

$$F^*(v^*;(x,v),t)=\partial_x H_F(x,v^*,t),\text{and } F_A(x,v^*,t)=\partial_{v^*}H_F(x,v^*,t)$$

respectively. Then it is easy to see that the result of corollary are equivalent with the conditions Eqs. (i) and (iv) of Theorem 3.1.

Below nonconvexity of a set-valued mapping $F(\cdot,t)$ means that its Hamilton function in general is a nonconcave function satisfying the condition Eq. (*a*).

Theorem 3.2. Suppose that we have the "nonconvex" problem (P_V), that is, $\varphi:(\mathbb{R}^n)^s\to\mathbb{R}^1$ and $F(\cdot,t)$: $\mathbb{R}^n\rightrightarrows\mathbb{R}^n$ in general are nonconvex function and set-valued mapping, respectively. Moreover, suppose that $K_{Q_j}\left(\tilde{x}^{(j)}(0)\right),\tilde{x}^{(j)}(0)\in Q_j$ is the cones of tangent directions to Q_j, $j=0,\ldots,s-1$.

Then for optimality of the trajectory $x(\cdot)$, it is sufficient that there exists an absolutely continuous function $x^*(\cdot)$, satisfying the following conditions:

(a) $L^*x^*(t)\in F^*(x^*(t);(\tilde{x}(t),L\tilde{x}(t)),t)$, a.e. $t\in[0,1]$,

(b) $\sum\limits_{k=0}^{s-1-j}(-1)^{s-k-j}D^{s-k-1-j}\left[p_{s-k}(0)x^*(0)\right]\in K^*_{Q_j}\left(\tilde{x}^{(j)}(0)\right),j=0,\ldots,s-1,$

(c) $\varphi(v_0,v_1,\ldots,v_{s-1})-\varphi\left(\tilde{x}(1),\tilde{x}'(1),\ldots,\tilde{x}^{(s-1)}(1)\right)$

$$\geq\sum_{j=0}^{s-1}\left\langle\sum_{k=0}^{s-1-j}(-1)^{s-k-j}D^{s-k-1-j}\left[p_{s-k}(1)x^*(1)\right],v_j-\tilde{x}^{(j)}(1)\right\rangle,$$

$$\forall v_j\in\mathbb{R}^n,j=0,\ldots,s-1,$$

(d) $\langle L\tilde{x}(t), x^*(t)\rangle = H_F(\tilde{x}(t), x^*(t), t)$, a.e. $t \in [0,1]$.

Proof. In the proof of Theorem 3.1, we have used the following inequality:

$$H_F(x(t), x^*(t), t) - H_F(\tilde{x}(t), x^*(t), t)$$
$$\leq \left\langle \sum_{k=1}^{s} (-1)^k \left(p_k(t)x^*(t)\right)^{(k)}, x(t) - \tilde{x}(t) \right\rangle. \tag{12}$$

Hence, from the inequality Eq. (12), immediately we have the inequality Eq. (10). Moreover, setting $v_j = \tilde{x}^{(j)}(1)\,(j = 0, \dots, s-1)$ for all feasible trajectories $x(\cdot)$, it is not hard to see that for nonconvex φ the following inequality holds:

$$\varphi\left(x(1), x'(1), \dots, x^{(s-1)}(1)\right) - \varphi\left(\tilde{x}(1), \tilde{x}'(1), \dots, \tilde{x}^{(s-1)}(1)\right)$$
$$\geq \sum_{j=0}^{s-1} \left\langle \sum_{k=0}^{s-1-j} (-1)^{s-k-j} D^{s-k-1-j}\left[p_{s-k}(1)x^*(1)\right], x^{(j)}(1) - \tilde{x}^{(j)}(1) \right\rangle, j = 0, 1, \dots, s-1,$$

Then for the furthest proof, we proceed by analogy with the preceding derivation of Theorem 3.1.

4. SOME APPLICATIONS TO OPTIMAL CONTROL PROBLEMS WITH PLDOS

In this section we give two applications of our results. The first one is the particular Mayer problem for differential inclusions involving PLDOs with constant coefficients, and the second one concerns optimization of "linear" differential inclusions with PLDOs and constant coefficients. Thus, suppose now we have the following optimization problem (for simplicity we consider a convex problem) with sth-order PLDO with constant coefficients:

$$\text{Minimize } \varphi_0(x(1)),$$
$$(P_C)\ Lx(t) \in F(x(t), t), \text{ a.e. } t \in [0,1], Lx = D^s x + p_1 D^{s-1} x + \dots + p_{s-1} Dx$$
$$x(0) = \alpha_0, x'(0) = \alpha_1, x''(0) = \alpha_2, \dots, x^{(s-1)}(0) = \alpha_{s-1}, . \tag{13}$$

where L is the sth-order polynomial operator, p_k, $k=1,\dots,s-1$ are some real constants, $F(\cdot,t): \mathbb{R}^n \rightrightarrows \mathbb{R}^n$ is a convex set-valued mapping, $\varphi_0: \mathbb{R}^n \to \mathbb{R}^1$ is a continuous convex function, and $\alpha_j \in \mathbb{R}^n, j=0,\dots,s-1$ are fixed n-dimensional vectors. It is known that the multiplication operation is commutative for polynomial linear differential operators with constant coefficients. On the other hand, the sth-order adjoint operator is defined as follows:

$$L^* x^* = (-1)^s D^s x^* + (-1)^{s-1} p_1 D^{s-1} x^* + \dots - p_{s-1} Dx^*.$$

Corollary 4.1. Let φ_0 and $F(\cdot,t)$ be a convex function and a set-valued mapping, respectively. Then, for the trajectory $x(\cdot)$ to be optimal in the problem (P_C), it is sufficient that there exists an absolutely continuous function $x^*(\cdot)$ satisfying the Euler-Lagrange type differential inclusion.

$$L^*x^*(t) \in F^*(x^*(t);(\tilde{x}(t),L\tilde{x}(t)),t);\ \langle L\tilde{x}(t),x^*(t)\rangle = H_F(\tilde{x}(t),x^*(t)),$$

$$\text{a.e. } t\in[0,1], L^*x^* = (-1)^s x^{*(s)} + (-1)^{s-1}p_1 x^{*(s-1)} + \cdots - p_{s-1}x^{*\prime}$$

and transversality condition at the endpoint t=1.

$$(-1)^s x^{*(s-1)}(1) \in \partial\varphi_0(\tilde{x}(1)), x^{*(j)}(1) = 0, j = 0, \ldots, m-2.$$

Proof. We conclude this proof by returning to the conditions Eqs. (i)–(iii) of Theorem 3.1. Clearly, a problem (P_C) can be reduced to the problem of form (P_V), where

$$\varphi\left(x(1),x'(1), \ldots, x^{(s-1)}(1)\right) \equiv \varphi_0(x(1)).$$

It follows that $\partial\varphi(x(1),x'(1), \ldots, x^{(s-1)}(1)) = \partial_x\varphi_0(x(1)) \times \underbrace{(0, \ldots, 0)}$. On the other hand, since $p_s(t)\equiv 1$, $p_j(t)$=p_s-j, and j=1,..., s-1 are constants, by sequentially substitution in the transversality condition Eq. (iii), we derive that

$$\sum_{k=0}^{s-1-j} (-1)^{s-k-j} D^{s-k-1-j}\left[p_{s-k}(1)x^*(1)\right] = x^{*(s-1-j)}(1) = 0, j = 1, \ldots, s-1,$$

and therefore for j=0.

$$\sum_{k=0}^{s-1} (-1)^{s-k} D^{s-k-1}\left[p_{s-k}(1)x^*(1)\right] = (-1)^s x^{*(s-1)}(1) \in \partial\varphi_0(\tilde{x}(1)).\square$$

Suppose now that we have the so-called linear Mayer problem with PLDOs:

$$\text{Minimize } \varphi_0(x(1)), \tag{14}$$

$$Lx(t) \in F(x(t),t), \text{ a.e. } t\in[0,1],$$

$$x^{(j)}(0) = \alpha_j, j = 0, \ldots, s-1, F(x,t) = A(t)x + B(t)U \tag{15}$$

where φ_0 differentiable convex function; $A(t)$ and $B(t)$ are $n\times n$ and $n\times r$ continuous matrices, respectively; U is a convex compact of $\mathbb{R}^r$; α_j, j=0,...,s-1 are constant vectors. It is required to find a control function $\tilde{u}(\cdot)$ such that the corresponding trajectory $\tilde{x}(\cdot)$ minimizes the Mayer functional $\varphi_0(x(1))$.

In fact, this is optimization of Cauchy problem for "linear" differential inclusions with PLDO. The controlling parameter $u(\cdot)$ is called admissible if it only takes values in the given control set U which is a nonempty, convex compact.

Theorem 4.1. The arc $x(t)$ corresponding to the controlling parameter $u(t)$ is a solution to Eqs. (14) and (15) if there exists an absolutely continuous function $x^*(\cdot)$, satisfying the Euler-Lagrange type differential equation, the transversality condition, and Weierstrass-Pontryagin maximum principle:

$$L^*x^*(t) = A^*(t)x^*(t), \text{a.e. } t \in [0,1],$$

$$(-1)^s x^{*(s-1)}(1) = \varphi_0'(\tilde{x}(1)), x^{*(j)}(1) = 0, j = 0, \ldots, s-2$$

$$\langle B(t)\tilde{u}(t), x^*(t)\rangle = \max_{u \in U} \langle B(t)u, x^*(t)\rangle.$$

Proof. Obviously, the Hamiltonian is

$$H_F(x, v^*, t) = \langle A(t)x, v^*\rangle + \max_{u \in U} \langle B(t)u, v^*\rangle.$$

Hence,

$$F^*(v^*; (x, \tilde{v}), t) = \partial_x H_F(x, v^*, t) = A^*(t)v^*, \tilde{v} \in F_A(x, v^*, t)\tilde{v} = A(t)x + B(t)\tilde{u}$$

where the argmaximum inclusion $\tilde{v} \in F_A(x, v^*, t)$implies that $\langle B(t)\tilde{u}, v^*\rangle = \max_{u \in U}\langle B(t)u, v^*\rangle$ $F^*(v^*; (x, \tilde{v}), t) \neq \varnothing$.Thus, applying Theorem 3.1, we obtain

$$L^*x^*(t) = A^*(t)x^*(t), \ L\tilde{x}(t) \in F_A(\tilde{x}(t), x^*(t), t),$$

$$\langle B(t)\tilde{u}(t), x^*(t)\rangle = \max_{u \in U} \langle B(t)u, x^*(t)\rangle.$$

Consequently, the transversality condition Eq. (ii) of Theorem 3.1 is unnecessary and by Corollary 4.1 $(-1)^s D^{s-1}x^*(1) = \varphi_0'(\tilde{x}(1))$,

Remark 4.1. Suppose that in the definition of *s*th-order PLDO (see Eq. (13)) *s*=1, p_i=0, *i*=1,..,*s*-1 and *U* is a convex closed polyhedron. Then we have linear equations with variable coefficients *x'*=*A*(*t*)*x*+*B*(*t*)*u*, *u*∈*U* in the finite time interval *t*∈[0,1]. Obviously, for such problems an adjoint Euler-Lagrange type differential equation and transversality condition at a point *t*=1 consist of the following: $x^{*\prime}(t) = -A^*(t)x^*(t), x^*(1) = -\varphi_0'(\tilde{x}(1))$. We remind that along with Pontryagin's maximum principle (see, e.g., [24]) under the condition for generality of position for time-optimal problem, the existence results of optimal control are proved.

Example 4.1. Let us consider the following Mayer problem with second-order PLDO $Lx=D^2x=x''$:

$$\text{Infimum } \varphi(x(1), x'(1)) \text{ is subject to } x'' = u, u \in [-1, 1], x(0) \in Q_0, x'(0) \in Q_1. \tag{16}$$

Here φ (x(1), x'(1)) = $x'^2(1)$ – x(1) and Q_0 = {0}, Q_1 = {1}.

It should be noted that substituting *F*(*t*)=*u*(*t*),*x''*(*t*)=*a*(*t*),*m*=1 into Newton's second law *F*(*t*)=*ma*(*t*), we have *x''*=*u*.

Obviously, in this problem $F(x,t) \equiv F(x) = \{u : |u| \leq 1\}, s = 2$.

Then Eq. (16) has the form:

$$\text{Infimum } \varphi(x(1), x'(1)) \text{ is subject to } Lx \in F(x), t \in [0,1], x(0) = 0, x'(0) = 1. \tag{17}$$

It can be easily seen that in the adjoint inclusion Eq. (i).

$$-D(p_1(t)x^*(t)) + D^2(p_2(t)x^*(t)) \in F^*(x^*(t); (\tilde{x}(t), L\tilde{x}(t)))$$

of Corollary 3.1 $p_2(t) \equiv 0$ and $p_2(t) \equiv 1$, and so we have

$$x^{*''}(t) \in F^*(x^*(t); (\tilde{x}(t), \tilde{x}''(t))).$$

Now, it is not hard to see that

$$H_F(x, v^*) = \max_u \{uv^* : |u| \leq 1\} = |v^*| \tag{18}$$

and

$$F^*(v^*; (x, v)) = \partial_x H_F(x, v^*) \equiv 0, \ v \in F_A(x, v^*) = \{-1, +1\}. \tag{19}$$

Then taking into account $Lx^* = d^2x^*/dt^2$, as a result of Theorem 3.1 (see also Corollary 3.1) from Eq. (19), we deduce that

$$x^{*''} = 0, \ t \in [0,1],$$

for which the solution is a linear function of the form $x^*(t) = C_1t + C_2$, where C_1, C_2 are arbitrary constants. Then Eq. (18) implies that $u(t)x^*(t) = |x^*(t)|$ or

$$\tilde{u}(t) = \begin{cases} \operatorname{sgn} x^*(t), & \text{if } x^*(t) \neq 0, \\ \forall u_0 \in [-1,1], & \text{if } x^*(t) = 0. \end{cases} \tag{20}$$

Further, from the linearity of $x^*(\cdot)$ and from Eq. (20), we insure that each optimal control function is a piecewise constant function.

In addition, by the transversality condition Eq. (iii) of Corollary 3.1, we can write.

$(x^{*'}(1), -x^*(1)) \in \partial\varphi(\tilde{x}(1), \tilde{x}'(1))$. On the other hand, it is not hard to see that $\varphi(x,y) = y^2 - x$ is a convex function; in fact, the 2×2 Hessian matrix

$$\varphi''(x,y) = \begin{bmatrix} \varphi''_{xx}(x,y) & \varphi''_{xy}(x,y) \\ \varphi''_{yx}(x,y) & \varphi''_{yy}(x,y) \end{bmatrix} = \begin{bmatrix} 0 & 0 \\ 0 & 2 \end{bmatrix}$$

is a positive semidefinite, that is, all eigenvalues of $\varphi''(x,y)$ are nonnegative. Indeed, denoting this matrix by A, we see that the characteristic equation $|A - \lambda E| = \lambda^2 - 2\lambda = 0$ (E is a 2×2 unique square matrix) has two real nonnegative eigenvalues $\lambda_1 = 0, \lambda_2 = 2$. Consequently, $\varphi(x,y)$ is convex and $\partial\varphi(x,y) = (-1, 2y)$. It follows that $\partial\varphi(\tilde{x}(1), \tilde{x}'(1)) = (-1, 2\tilde{x}'(1))$. Comparing this relation with $(x^{*'}(1), -x^*(1)) \in \partial\varphi(\tilde{x}(1), \tilde{x}'(1))$, we immediately have $x^{*'}(1) = -1, x^*(1) = -2x'(1)$.. Then from a general solution of the adjoint Euler-Lagrange type inclusion (equation) $x^*(t) = C_1t + C_2$, we have $-2\tilde{x}'(1) = x^*(1) = C_1 + C_2, \ -1 = x^{*'}(1) = \ C_1 \ (C_1, C_2$

are arbitrary constants), and so $x^*(t) = 1 - t - 2\tilde{x}'(1)$, whence $x^*(t) \neq 0$, if $t \neq \tau = 1 - 2\tilde{x}'(1)$. Therefore, Eq. (20) implies that for optimal control $u(\cdot)$, there are four possibilities:

$$\tilde{u}(t) = 1,\ x^*(t) > 0,\ t \in [0,1]. \tag{21}$$

$$\tilde{u}(t) = -1,\ x^*(t) < 0, t \in [0,1]. \tag{22}$$

$$\tilde{u}(t) = \begin{cases} 1, & \text{if} \quad 0 \le t < \tau, \\ -1, & \text{if} \quad \tau < t \le 1. \end{cases} \tag{23}$$

$$\tilde{u}(t) = \begin{cases} -1, & \text{if} \quad 0 \le t < \tau, \\ 1, & \text{if} \quad \tau < t \le 1. \end{cases} \tag{24}$$

(observe that τ is a point of discontinuity of $\tilde{u}(\cdot)$ and the values of the control functions $\tilde{u}(\cdot)$ at a point of discontinuity τ are unessential). As a consequence, it follows that either the sign of the linear function $x^*(t)$ does not change for the whole interval [0,1] or $x^*(t)>0$, $0\le t<\tau$; $x*(t)<0$, $\tau<t\le 1$ for a some τ in the interval $0<\tau<1$ (the case Eq. (24) is excluded). Therefore, since $u(t)$ is a piecewise constant function, having not more than two intervals of constancy, we have either the cases Eqs. (21) and (22) or the case Eq. (23). In general, using Eqs. (21)–(23), by solving the Cauchy problem

$$x''(t) = u(t),\ x(0) = 0, x'(0) = 1 \tag{25}$$

we have a unique solution of the initial value problem Eq. (25). Thus for the time interval on which u=1, we have $x'(t)=t+c_1; x(t)=t^2/2+c_1t+c_2$ (c_1, c_2 are constants). From Eq. (25) we obtain.

$$x'(t) = t + 1; x(t) = t^2/2 + t. \tag{26}$$

By analogy, for u=-1 we have.

$$x'(t) = 1 - t;\ x(t) = -t^2/2 + t. \tag{27}$$

Now, let $x_1(t)$ and $x_2(t)$ be parabolas of Eqs. (26) and (27), respectively. Here, in the case Eq. (21), $u(t)=1, t\in[0,1]$, and so from Eq. (26), we have $\tilde{x}_1(1) = 0.5 + 1 = 1.5; \tilde{x}_1'(1) = 2.$; Consequently, the value of problem Eq. (16) is $\varphi(\tilde{x}_1(1), \tilde{x}_1'(1)) = \tilde{x}_1'^2(1) - \tilde{x}_1(1) = 2^2 - (1.5) = 2.5$, if $\tilde{u}(t) = 1, t \in [0,1]$. By a similar way, for a control function $\tilde{u}(t) = -1, t \in [0,1]$ from Eq. (27), we obtain that $\tilde{x}_2(1) = -1^2/2 + 1 = 0.5, \tilde{x}_2'(1) = 0$ and $\varphi(\tilde{x}_2(1), \tilde{x}_2'(1)) = 0^2 - (0.5) = -0.5$.

On the other hand, in the case Eq. (23), the control function $\tilde{u}(t)$ first is equal to +1 and then equal to -1, and the trajectory $\tilde{x}(t)$ consists of two pieces of parabolas $\tilde{x}_1(t)$ and $\tilde{x}_2(t)$ ($\tilde{x}(t)$ is continuous and piecewise smooth on the interval 0≤t≤1. Then the solution of the equation Eq. (25) on the interval 0≤t≤τ is given by Eq. (26); at a point τ are satisfied $x_1(\tau) = \tau^2/2 + \tau$, $x_1'(\tau) = 1 + \tau$.. Consider now the initial value problem:

$$x_2''(t) = -1, \ x_2(\tau) = \tau^2/2 + \tau, x_2'(\tau) = 1 + \tau, t \in [\tau, 1]. \tag{28}$$

It is clear that $(\tau^2/2) + \tau = x_2(\tau) = -(\tau^2/2) + c_1\tau + c_2$ and $1 + \tau = x_2'(\tau) = -\tau + c_1$ from which we obtain that the solution of the initial value problem Eq. (28) is $\tilde{x}_2(t) = -(t^2/2) + (1 + 2\tau)t - \tau^2$. Substituting the value $\tau = 1 - 2\tilde{x}'(1)$ into equation $\tilde{x}_2'(t) = 1 - t + 2\tau$, we have $5\tilde{x}_2'(1) = 2, \tilde{x}_2'(1) = \tilde{x}'(1) = 0.4$ (it follows that τ=0.2). Moreover, $\tilde{x}_2(1) = 2\tau - \tau^2 + (0.5)$ and $\tilde{x}_2(1) = \tilde{x}(1) = (3/2) - 4\tilde{x}'^2(1) = 0.86$.

Thus, the value of our Mayer problem is $\varphi(\tilde{x}(1), \tilde{x}'(1)) = \tilde{x}'^2(1) - \tilde{x}(1) = (2/5)^2 - (43/50) = -0.7$, where *u* (*t*) is defined as in Eq. (23). Comparing the values 2.5,-0.5,-0.7, we believe that the value of Mayer problem is -0.7.

5. SUFFICIENT CONDITIONS OF OPTIMALITY FOR SECOND-ORDER EVOLUTION DIFFERENTIAL INCLUSIONS WITH ENDPOINT CONSTRAINTS

Note that in this section the optimality conditions are given for second-order convex differential inclusions (P_M) with convex endpoint constraints. These conditions are more precise than any previously published ones since they involve useful forms of the Weierstrass-Pontryagin condition and second-order Euler-Lagrange type adjoint inclusions. In the reviewed results, this effort culminates in Theorem 5.1:

$$\text{Minimize}\, g(x(1), x'(1)),$$

$$(\mathrm{P_M})\, x''(t) \in F(x(t), x'(t), t), \text{ a.e. } t \in [0, 1],$$

$$x(0) = x_0, x'(0) = x_1; x(1) \in M_0, \ x'(1) \in M_1,$$

where *g* is a convex continuous function, $F(\cdot,t)\colon \mathbb{R}^{2n} \rightrightarrows \mathbb{R}^n$ is convex set-valued mapping, and $M_0, M_1 \subseteq \mathbb{R}^n$ are convex sets.

The following adjoint inclusion is the second-order Euler-Lagrange type inclusion for the problem (P_M):

$$\mathrm{a}_1.\ (x^{*\prime\prime}(t) + v^{*\prime}(t), v^*(t)) \in F^*(x^*(t); (\tilde{x}(t), \tilde{x}'(t), \tilde{x}''(t)), t), \text{ a.e. } t \in [0, 1],$$

where

$$\mathrm{b}_1.\ \tilde{x}''(t) \in F_A(\tilde{x}(t), \tilde{x}'(t); x^*(t), t), \text{ a.e. } t \in [0, 1].$$

In what follows we assume that $x^*(t), t \in [0,1]$ is absolutely continuous function together with the first-order derivatives for which $x^{*\prime\prime}(\cdot) \in L_1^n([0, 1])$. Besides the auxiliary function $v^*(t), t \in [0, 1]$ is absolutely continuous and $v^{*\prime}(\cdot) \in L_1^n([0, 1])$.

The transversality conditions at the endpoint t=1 consist of the following:

c_1. $(v^*(1)+x^{*\prime}(1), -x^*(1)) \in \partial_{(x,u)} g(\tilde{x}(1),\tilde{x}'(1)) - K^*_{M_0}(\tilde{x}(1)) \times K^*_{M_1}(\tilde{x}'(1))$.

Now we are ready to formulate the following theorem of optimality.

Theorem 5.1. Suppose that g is a continuous and convex function, $F(\cdot,t)$ is a convex set-valued mapping, and M_0, M_1 are convex sets. Then for optimality of the feasible trajectory $\tilde{x}(t)$ in the problem (P_M), it is sufficient that there exists a pair of absolutely continuous functions:

$$\{x^*(t), v^*(t)\}, t \in [0,1]$$

satisfying a.e. the second-order Euler-Lagrange type inclusions Eqs. (a_1) and (b_1) and the transversality condition Eq. (c_1) at the endpoint t=1.

Proof. By the proof idea of Theorem 3.1 from Eqs. (a_1) and (b_1), we obtain the adjoint differential inclusion of second order:

$$(x^{*\prime\prime}(t)+v^{*\prime}(t), v^*(t)) \in \partial_{(x,u)} H_F(\tilde{x}(t),\tilde{x}'(t),x^*(t),t), t \in [0,1].$$

On the definition of subdifferential set of the Hamiltonian function $H_F(\cdot,t)$ for all feasible trajectory $x(t)$,$t\in[0,1]$ we rewrite the last relation in the equivalent form:

$$H_F(x(t),x'(t),x^*(t),t) - H_F(\tilde{x}(t),\tilde{x}'(t),x^*(t),t)$$

$$\leq \langle x^{*\prime\prime}(t)+v^{*\prime}(t), x(t)-\tilde{x}(t)\rangle + \langle v^*(t), x'(t)-\tilde{x}'(t)\rangle. \tag{29}$$

Now by using definition of the Hamiltonian function, the inequality Eq. (29) can be reduced to the inequality

$$0 \geq \langle (x(t)-\tilde{x}(t))'', x^*(t)\rangle - \langle x^{*\prime\prime}(t), x(t)-\tilde{x}(t)\rangle - \frac{d}{dt}\langle v^*(t), x(t)-\tilde{x}(t)\rangle. \tag{30}$$

Integrating of the inequality Eq. (30) over the interval [0,1], we derive that

$$0 \geq \int_0^1 [\langle (x(t)-\tilde{x}(t))'', x^*(t)\rangle - \langle x^{*\prime\prime}(t), x(t)-\tilde{x}(t)\rangle] dt$$

$$+\langle v^*(0), x(0)-\tilde{x}(0)\rangle - \langle v^*(1), x(1)-\tilde{x}(1)\rangle. \tag{31}$$

For convenience we transform the expression in the square parentheses on the right-hand side of Eq. (31) as follows

$$\langle (x(t)-\tilde{x}(t))'', x^*(t)\rangle - \langle x^{*\prime\prime}(t), x(t)-\tilde{x}(t)\rangle$$

$$= \frac{d}{dt}\langle (x(t)-\tilde{x}(t))', x^*(t)\rangle - \frac{d}{dt}\langle x^{*\prime}(t), x(t)-\tilde{x}(t)\rangle.$$

Thus by elementary property of the definite integrals, we can compute the integral on the right-hand side of Eq. (31):

$$\int_0^1 \left[\langle (x(t) - \tilde{x}(t))'', x^*(t) \rangle - \langle x^{*''}(t), x(t) - \tilde{x}(t) \rangle \right] dt$$

$$\begin{aligned} &= \langle x'(1) - \tilde{x}'(1), x^*(1) \rangle - \langle x'(0) - \tilde{x}'(0), x^*(0) \rangle \\ &- \langle x^{*'}(1), x(1) - \tilde{x}(1) \rangle + \langle x^{*'}(0), x(0) - \tilde{x}(0) \rangle. \end{aligned} \tag{32}$$

Then substituting Eq. (32) into Eq. (31), we have

$$\begin{aligned} 0 &\geq \langle x'(1) - \tilde{x}'(1), x^*(1) \rangle - \langle x'(0) - \tilde{x}'(0), x^*(0) \rangle \\ &- \langle v^*(1) + x^{*'}(1), x(1) - \tilde{x}(1) \rangle + \langle v^*(0) + x^{*'}(0), x(0) - \tilde{x}(0) \rangle. \end{aligned} \tag{33}$$

Now, remember that $x(\cdot), \tilde{x}(\cdot)$ are feasible trajectories and $x(0) = \tilde{x}(0) = x_0$ and $x'(0) = \tilde{x}'(0). \ = x_1$. Then it follows from Eq. (33) that

$$0 \geq \langle x'(1) - \tilde{x}'(1), x^*(1) \rangle - \langle v^*(1) + x^{*'}(1), x(1) - \tilde{x}(1) \rangle. \tag{34}$$

Now, thanking to the transversality conditions Eq. (c_1) at the endpoint t=1, we can rewrite

$$\begin{aligned} &g(x(1), x'(1)) - g(\tilde{x}(1), \tilde{x}'(1)) \geq \langle v^*(1) + x^{*'}(1) + x^*(1), x(1) - \tilde{x}(1) \rangle \\ &+ \langle x^{*'}(1) - x^*(1), x'(1) - \tilde{x}'(1) \rangle, \ x^*(1) \in K^*_{M_0}(\tilde{x}(1)), \ x^{*'}(1) \in K^*_{M_1}(\tilde{x}'(1)) \end{aligned}$$

or, in other words

$$\begin{aligned} &g(x(1), x'(1)) - g(\tilde{x}(1), \tilde{x}'(1)) \\ &\geq \langle v^*(1) + x^{*'}(1), x(1) - \tilde{x}(1) \rangle - \langle x^*(1), x'(1) - \tilde{x}'(1) \rangle \end{aligned} \tag{35}$$

Thus, summing the inequalities Eqs. (34) and (35) for all feasible trajectories $x(\cdot)$, satisfying the initial conditions $x(0)=x_0$ and $x'(0)=x_1$ and endpoint constraints $x(1) \in M_0, x'(1) \in M_1$, we have the needed inequality:

$$g(x(1), x'(1)) - g(\tilde{x}(1), \tilde{x}'(1)) \geq 0 \text{ or } g(x(1), x'(1)) \geq g(\tilde{x}(1), \tilde{x}'(1)). \ \square$$

6. CONCISE SUMMARY

The suggested method investigates the problem of differential inclusions described by polynomial linear differential operators. This problem is a significant extension of problems involving first-order differential inclusions. Therefore, the necessary conditions for optimal solutions to such situations are derived. The presence of nonfunctional initial point or endpoint constraints gives rise to several types of transversality criteria. In addition, it is undeniable that studying the optimality conditions of problems involving second- and fourth-order Sturm-Liouville type differential inclusions can have a significant impact on the advancement of contemporary optimization. Furthermore, there is strong evidence to suggest that this impact will become even more substantial in the coming years. Therefore, the proposed issue with linear differential operators with variable coefficients might be employed in numerous ways to solve practical situations.

7. AT A GLANCE

This chapter focused on a novel category of problems in the field of optimal control theory, which are characterized by polynomial linear differential operators. Consequently, a fascinating Mayer problem develops with higher order differential inclusions. Therefore, the formulation of sufficient optimality criteria is achieved by considering the Euler-Lagrange and Hamiltonian type inclusions. Furthermore, the establishment of transversality conditions at the boundaries of the temporal interval being considered is crucial for future investigations. In order to achieve this objective, the apparatus of locally adjoint mappings is employed, which is crucial in the primary findings of this chapter. The method is illustrated using the linear optimal control problem as an example, where the Weierstrass-Pontryagin maximum principle is developed.

REFERENCES

1. Auslender A, Mechler J. Second order viability problems for differential inclusions.Journal of Mathematical Analysis and Applications. 1994;181:205-218
2. Azzam-Laouir D, Castaing C, Thibault L. Three boundary value problems for second order differential inclusion in Banach spaces. Control and Cybernetics. 2002;31:659-693
3. Cernea A. On the existence of viable solutions for a class of second order differential inclusions. Discussiones Mathematicae. Differential Inclusions, Control and Optimization. 2002;22:67-78
4. Barbu V, Precupanu T. Convex Control Problems in Banach Spaces. 4th ed. Netherlands: Springer; 2012. 368 p
5. Blagodatskikh VI, Filippov AF. Differential inclusions and optimal control. Trudy Matematicheskogo Instituta Imeni VA Steklova. 1985;169:194-252
6. Buttazzo G, Drakhlin ME, Freddi L, Stepanov E. Homogenization of optimal control problems for functional differential equations. Journal of Optimization Theory and Applications. 1997;93:103-119
7. Cannarsa P, Sinestrari C. Semiconcave Functions, Hamilton-Jacobi Equations and Optimal Control. Boston: Birkhäuser; 2004. 304p
8. Clarke FH. Functional analysis, calculus of variations and optimal control. Graduate Texts in Mathematics. London: Springer; 2013. 264p
9. Kolmogorov AN, Fomin SV. Elements of the Theory of Functions and Functional Analysis. (Dover Books on Mathematics) Paperback. New York: Dover Publications; 1999. 128p

10. Lasiecka I, Triggiani R. Control Theory for Partial Differential Equations: Vol. 1, Abstract Parabolic Systems: Continuous and Approximation Theories. Cambridge: Cambridge University Press; 2000. 672p
11. Papageorgiou NS, Rădulescu VD. Periodic solutions for time-dependent subdifferential evolution inclusions. Evolution Equations & Control Theory. 2017;6:277-297
12. Long DQ, Dang Quang A, Luan VT. Iterative method for solving a fourth order differential equation with nonlinear boundary condition. Applied Mathematical Sciences. 2010;4:3467-3481
13. Zhou Y, Vijayakumar V, Murugesu R. Controllability for fractional evolution inclusions without compactness. Evolution Equations and Control Theory. 2015;4:507-524
14. Mahmudov EN. Optimization of Mayer problem with Sturm–Liouville-type differential inclusions. Journal of Optimization Theory and Applications. 2018;177:345-375
15. Mahmudov EN. Approximation and optimization of higher order discrete and differential inclusions. Nonlinear Differential Equations and Applications NoDEA. 2014;21:1-26
16. Mahmudov EN. Convex optimization of second order discrete and differential inclusions with inequality constraints. Journal of Convex Analysis. 2018;25:1-26
17. Mahmudov EN. Mathematical programming and polyhedral optimization of second order discrete and differential inclusions. Pacific Journal of Optimization. 2015;11:495-525
18. Mahmudov EN. Free time optimization of higher order differential inclusions with endpoint constraints. Applicable Analysis. 2018;97:2071-2084
19. Mahmudov EN. Optimization of boundary value problems for certain higher-order differential inclusions. Journal of Dynamical and Control Systems. 2019;25:17-27
20. Mahmudov EN. Optimization of second order differential inclusions with boundary value conditions. Journal of Nonlinear and Convex Analysis. 2017;18:1653-1664
21. Mahmudov EN. Optimal control of second order delay-discrete and delay differential inclusions with state constraints. Evolution Equations & Control Theory. 2018;7:501-529. DOI: 10.3934/eect.2018024

22. Bors D, Majewski M. On Mayer problem for systems governed by second-order ODE. Optimization. 2014;63:239-254
23. Mahmudov EN. Approximation and Optimization of Discrete and Differential Inclusions. Waltham, USA: Elsevier; 2011. 396p
24. Pontryagin LS, Boltyanskii VG, Gamkrelidze RV, Mishchenko EF. The Mathematical Theory of Optimal Processes. New York/London/Sydney: John Wiley & Sons, Inc.; 1965. 360p
25. Mahmudov EN. Optimal control of Cauchy problem for first order discrete and partial differential inclusions. Journal of Dynamical and Control Systems. 2009;15:587-610
26. Mordukhovich BS. Variational Analysis and Generalized Differentiation, I: Basic Theory. II: Applications, Grundlehren Series (Fundamental Principles of Mathematical Sciences). Vol. 330 and 331. New York: Springer; 2006. 579p
27. Mahmudov EN. Necessary and sufficient conditions for discrete and differential inclusions of elliptic type. Journal of Mathematical Analysis and Applications. 2006;323:768-789.

09 ANALYZE THE SPECTRAL PROPERTIES AND CONDUCT NUMERICAL INVESTIGATIONS ON A FLEXIBLE STRUCTURE WITH NONCONSERVATIVE BOUNDARY CONDITIONS

1. INTRODUCTION: AN OVERVIEW

This work focuses on the spectral analysis and numerical investigation of the eigenvalues of the Euler-Bernoulli beam model. The left end of the beam is clamped and the beam is subjected to linear feedback conditions with a non-dissipative feedback matrix [1, 2]. The model's behavior can vary from being conservative, dissipative, or fully non-dissipative, depending on the values of the boundary parameters k1 and k2. Our focus is on the non-dissipative scenario, where the energy of a vibrating system does not decrease or stay the same over time. Our technique involves simplifying the initial-boundary value problem that describes the beam dynamics to a first-order equation in the state Hilbert space $\mathcal{H}$.The system's development is solely dictated by the dynamics generator, denoted as $\mathcal{L}_{k_1,k_2}$, which is an unbounded non-self-adjoint matrix differential operator. Please refer to Equations (2), (3), and (8) for further details. The eigenmodes and mode forms of the flexible structure are determined by the eigenvalues (multiplied by a constant i) and the generalized eigenvectors of $\mathcal{L}_{k_1,k_2}$.

Based on the results of [1, 2], the dynamics generator has a purely discrete spectrum, whose location on the complex plane is determined by the controls k_1 and k_2. Having in mind the practical applications of the asymptotic formulas [3, 4, 5], we discuss the case of $k_1 \geq 0$ and $k_2 \geq 0$, such that $|k_1|+|k_2|>0$ (see Proposition 2). As shown in [2], even though the operator $\mathcal{L}_{k1,k2}$ is non-dissipative, for the case $k_1>0$ and $k_2=0$ (or $k_1=0$ and $k_2>0$), the entire set of eigenvalues is located in the closed upper half of the complex plane $\mathbb{C}$, which means that all eigenmodes are stable or neutrally stable. (We recall that to obtain an elastic mode from an eigenvalues of $\mathcal{L}_{k1,k2}$, one should multiply the eigenvalue by a factor *i*).

In the paper we address the question of accuracy of the asymptotic formulas for the eigenvalues. *Namely, under what conditions the leading asymptotic terms in*formulas (20) *and* (21) *can be used for practical estimation of the actual frequencies of the flexible beam?* Numerical simulations show that the accuracy of the asymptotic formulas is really high; the leading asymptotic terms can be used by practitioners almost immediately, i.e., almost from the first vibrational mode. The second question is concerned with the role of the *deadbeat modes*. A deadbeat mode is a purely negative elastic mode that generates a solution of the evolution equation exponentially decaying in time. The deadbeat modes are important in engineering applications. As we prove in the paper, when the boundary parameter k_1 is close to 1 (while k_2=0), the number of the deadbeat modes is so large that the corresponding mode shapes become important for the description of the beam dynamics. More precisely, the number of deadbeat modes tends to infinity as $k_1 \to 1^+$.

We have also shown that there exists a sequence of values of the parameter k_1, i.e., $\left\{k_1^{(n)}\right\}_{n=1}^{\infty}$, such that for each $k_1 = k_1^{(n)}$ there exist a finite number of deadbeat modes and each corresponds to a *double eigenvalue* of the dynamics generator $\mathcal{L}_{k1,k2}$. For each value $k_1^{(n)}$, the operator $\mathcal{L}k_{1,k2}$ has a two-dimensional root subspace spanned by an eigenvector and an associate vector. This result means that for a double deadbeat mode (corresponding to $k_1^{(n)}$), there exists a mode shape and an associate mode shape. This fact indicates that for some values of k_1 and k_2, there exists a significant number of associate vectors of $\mathcal{L}_{k1,k2}$. Therefore, if one can prove that the set of the generalized eigenvectors (eigenvectors and associate vectors together) forms an unconditional basis for the state space, then construction of the bi-orthogonal basis [6] would be a more complicated problem than for the case when no associate vectors exist.

Finally, we mention that the feedback control of beams is a well-studied area [6], with multiple applications to the control of robotic manipulators, long and slender aircraft wings, propeller blades, large space structure [7, 8], and the dynamics of carbon nanotubes [9]. The analysis of a classical beam model with nonstandard feedback control law that originated in engineering literature [4, 10, 11, 12] may be of interest for both analysts and practitioners.

The structure of this document is as follows. Section 1 presents the formulation of the initial-boundary value problem for the Euler-Bernoulli beam model. Section 2 presents a reformulation of the problem as an evolution equation in the Hilbert space of Cauchy data, also known as the energy space. The primary focus is on the dynamics generator, denoted as $\mathcal{L}_{k_1,k_2}$, which is a matrix differential operator that is not self-adjoint. This operator depends on two parameters, k1 and k2. The eigenvalues and generalized eigenvectors of λk1,k2 correspond to the modes and mode shapes of the beam. In addition, we provide numerical approximations and graphical representations of the eigenvalues of a discrete approximation of the primary operator. Please refer to Tables 1 and 2 and Figures 1 and 2 for further details. Section 3 focuses on analyzing the deadbeat modes and determining the lower and upper bounds for the number of these modes. This is done for various values of the boundary parameters, as illustrated in Figure 5. Section 4 focuses on the asymptotic approximation of the collection of double deadbeat modes. This information can be found in Tables 3 and 4, as well as Figures 6 and 7. Section 5 provides a detailed description of the numerical method employed to analyze the spectral properties of the finite-dimensional approximation of the dynamics generator.

Table 1. Approximations of the eigenvalues for the discrete and "continuous" operators (*K*>1).

k_1=0, k_2=0.5, *EI*=1 EI=1, ρ=0.1, *N*=64, ε*f*=10^{-20}					
No.	Numerical	Analytic	No.	Numerical	Analytic
1.	-7988.1+237.00*i*	-7988.1+237.00*i*	18.	28.860+13.515*i*	29.453+14.812*i*
2.	-7020.6+222.19*i*	-7020.6+222.19*i*	19.	123.16+29.723*i*	123.08+29.625*i*
3.	-6115.5+207.37*i*	-6115.5+207.37*i*	20.	279.13+44.431*i*	279.14+44.437*i*
4.	-5272.8+192.56*i*	-5272.8+192.56*i*	21.	497.61+59.250*i*	497.61+59.250*i*
5.	-4492.5+177.75*i*	-4492.5+177.75*i*	22.	778.50+74.062*i*	778.50+74.062*i*
6.	-3774.7+162.94*i*	-3774.7+162.94*i*	23.	1121.8+88.875*i*	1121.8+88.875*i*
7.	-3119.3+148.12*i*	-3119.3+148.12*i*	24.	1527.6+103.69*i*	1527.6+103.69*i*
8.	-2526.3+133.31*i*	-2526.3+133.31*i*	25.	1995.7+118.50*i*	1995.7+118.50*i*
9.	-1995.7+118.50*i*	-1995.7+118.50*i*	26.	2526.3+133.31*i*	2526.3+133.31*i*
10.	-1527.6+103.69*i*	-1527.6+103.69*i*	27.	3119.3+148.12*i*	3119.3+148.12*i*
11.	-1121.8+88.875*i*	-1121.8+88.875*i*	28.	3774.7+162.94*i*	3774.7+162.94*i*
12.	-778.5+74.062*i*	-778.5+74.062*i*	29.	4492.5+177.75*i*	4492.5+177.75*i*
13.	-497.61+59.250*i*	-497.61+59.250*i*	30.	5272.8+192.56*i*	5272.8+192.56*i*
14.	-279.13+44.431*i*	-279.14+44.437*i*	31.	6115.5+207.37*i*	6115.5+207.37*i*
15.	-123.16+29.723*i*	-123.08+29.625*i*	32.	7020.6+222.19*i*	7020.6+222.19*i*
16.	-28.860+13.515*i*	-29.453+14.812*i*	33.	7988.1+237.00*i*	7988.1+237.00*i*
17.	-2.2·10^{-17}+4.6007*i*				

Table 2. Approximations of the eigenvalues for the discrete and "continuous" operators (*K*<-1).

k_1=1.3, k_2=1.2, *EI*=10 *EI*=10, ρ=0.1, *N*=64, ε*f*=10^{-20}					
No.	Numerical	Analytic	No.	Numerical	Analytic
1.	-25266-229.07*i*	-25266-229.07*i*	18.	99.467-19.816*i*	98.177-14.317*i*
2.	-22206-214.76*i*	-22206-214.76*i*	19.	394.17-28.149*i*	394.26-28.634*i*
3.	-19344-200.44*i*	-19344-200.44*i*	20.	887.75-42.983*i*	887.75-42.951*i*
4.	-16679-186.12*i*	-16679-186.12*i*	21.	1578.6-57.267*i*	1578.6-57.268*i*
5.	-14212-171.81*i*	-14212-171.81*i*	22.	2466.9-71.586*i*	2466.9-71.585*i*
6.	-11942-157.49*i*	-11942-157.49*i*	23.	3552.5-85.902*i*	3552.5-85.903*i*
7.	-9869.1-143.17*i*	-9869.1-143.17*i*	24.	4835.6-100.22*i*	4835.6-100.22*i*
8.	-7993.9-128.85*i*	-7993.9-128.85*i*	25.	6316.0-114.54*i*	6316.0-114.54*i*
9.	-6316.0-114.54*i*	-6316.0-114.54*i*	26.	7993.9-128.85*i*	7993.9-128.85*i*
10.	-4835.6-100.22*i*	-4835.6-100.22*i*	27.	9869.1-143.17*i*	9869.1-143.17*i*
11.	-3552.5-85.902*i*	-3552.5-85.903*i*	28.	11942-157.49*i*	11942-157.49*i*
12.	-2466.9-71.586*i*	-2466.9-71.585*i*	29.	14212-171.81*i*	14212-171.81*i*
13.	-1578.6-57.267*i*	-1578.6-57.268*i*	30.	16679-186.12*i*	16679-186.12*i*
14.	-887.75-42.983*i*	-887.75-42.951*i*	31.	19344-200.44*i*	19344-200.44*i*
15.	-394.17-28.149*i*	-394.26-28.634*i*	32.	22206-214.76*i*	22206-214.76*i*
16.	-99.467-19.816*i*	-98.177-14.317*i*	33.	25266-229.07*i*	25266-229.07*i*
17.	1.5·10^{-17}+7.7256*i*				

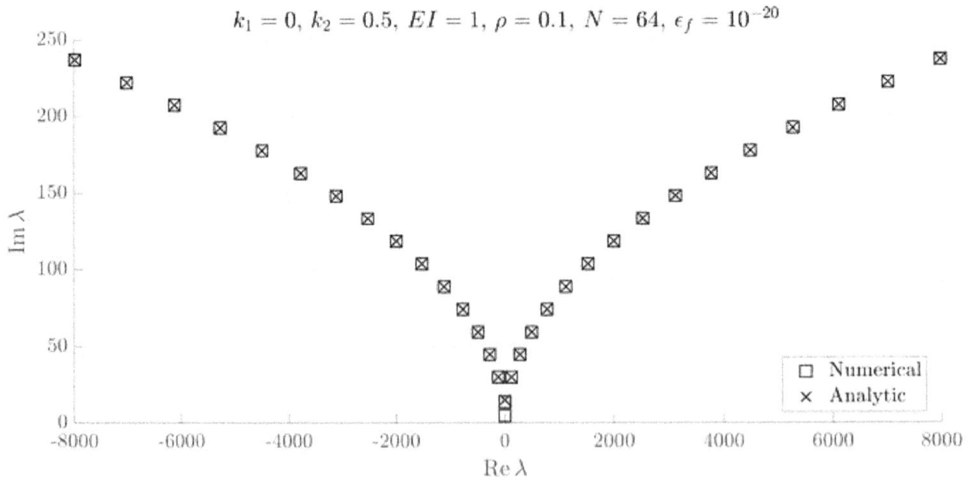

Figure 1. Graphical representation of the eigenvalues of the discrete and "continuous" operators (*K*>1).

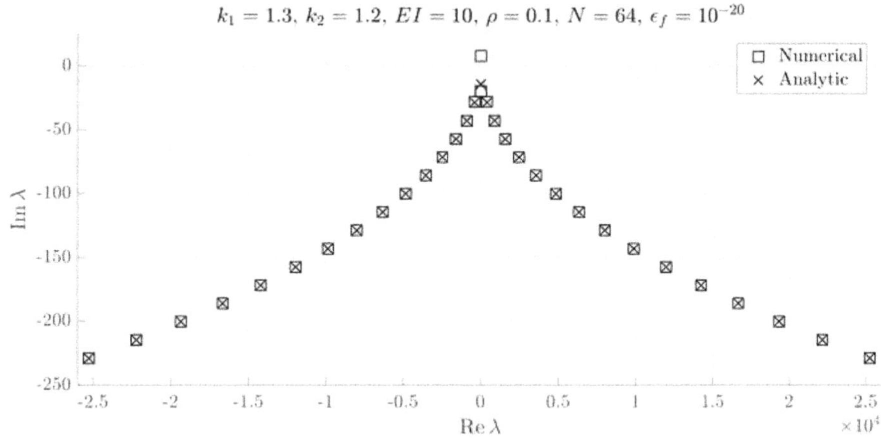

Figure 2. Graphical representation of the eigenvalues of the discrete and "continuous" operators ($K<-1$).

Table 3. Eigenvalues closest to the imaginary axis as $k_1 \to 1^+$.

k_2=0, EI=1 *EI*=1, ρ=1, *N*=64, *εf*=10^{-30}			
No.	$k_1=1+10^{-4}$	$k_1=1+10^{-7}$	$k_1=1+10^{-10}$
1.	$-222.22+155.56i$	$-176.06+264.07i$	$-106.90+372.41i$
2.	$-133.38+124.45i$	$-87.723+211.27i$	$-2.4378\cdot10^{-16}+254.44i$
3.	$-64.540+93.396i$	$-3.9609\cdot10^{-18}+162.37i$	$-3.0717\cdot10^{-17}+123.66i$
4.	$-9.8081+58.559i$	$-5.7977\cdot10^{-19}+116.23i$	$-1.3012\cdot10^{-17}+4.9349i$
5.	$-7.7725\cdot10^{-21}+5.0488i$	$-6.3661\cdot10^{-20}+44.182i$	$-8.9232\cdot10^{-18}+44.421i$
6.	$4.9365\cdot10^{-21}+38.994i$	$-6.0066\cdot10^{-20}+4.9383i$	$8.9143\cdot10^{-18}+44.406i$
7.	$7.8007\cdot10^{-21}+4.8257i$	$6.0114\cdot10^{-20}+4.9313i$	$1.3012\cdot10^{-17}+4.9347i$
8.	$9.8081+58.559i$	$6.6801\cdot10^{-20}+44.651i$	$2.9907\cdot10^{-17}+123.09i$
9.	$64.540+93.396i$	$2.9381\cdot10^{-18}+139.20i$	$1.1286\cdot10^{-16}+234.06i$
10.	$133.38+124.45i$	$87.723+211.27i$	$3.2280\cdot10^{-16}+315.13i$
11.	$222.22+155.56i$	$176.06+264.07i$	$106.90+372.41i$

Table 4. Eigenvalues closest to the imaginary axis as $k_1 \to 1^-$.

k_2=0, *EI*=1 *EI*=1, ρ=1, *N*=64, *εf*=10^{-30}			
#	$k_1=1-10^{-4}$	$k_1=1-10^{-7}$	$k_1=1-10^{-10}$
1.	$-175.34+140.01i$	$-129.20+237.56i$	$-60.143+338.80i$
2.	$-96.394+108.84i$	$-50.206+186.90i$	$-8.6602+240.01i$
3.	$-36.896+78.778i$	$-8.0431+121.15i$	$-0.28608+123.37i$
4.	$-6.0769+42.171i$	$-0.23455+44.410i$	$-0.0074192+44.413i$
5.	$-0.11141+4.9324i$	$-0.0035253+4.9348i$	$-0.00011148+4.9348i$
6.	$0.11141+4.9324i$	$0.0035253+4.9348i$	$0.00011148+4.9348i$
7.	$6.0769+42.171i$	$0.23455+44.410i$	$0.0074192+44.413i$

k_2=0, EI=1 EI=1, ρ=1, N=64, εf=10^{-30}			
8.	36.896+78.778i	8.0431+121.15i	0.28608+123.37i
9.	96.394+108.84i	50.206+186.90i	8.6602+240.01i
10.	175.34+140.01i	129.20+237.56i	60.143+338.80i

1.1 The initial-boundary value problem for the Euler-Bernoulli beam model of a unit length

The Lagrangian of the system is defined by [10, 11]

$$\frac{1}{2}\int_0^1 \left[\varrho(x)A(x)h_t^2(x,t) - E(x)I(x)h_{xx}^2(x,t)\right]dx, \tag{1}$$

where $h(x,t)$ is the transverse deflection, $E(x)$ is the modulus of elasticity, $I(x)$ is the area moment of inertia, $\varrho(x)$ is the linear density, and $A(x)$ is the cross-sectional area of the beam.

Assuming that the beam is clamped at the left end (x=0) and free at the right end (x=1), and applying Hamilton's variational principle to the action functional defined by (1), we obtain the equation of motion

$$\varrho(x)A(x)h_{tt}(x,t) + (E(x)I(x)h_{xx}(x,t))_{xx} = 0, \qquad 0 \le x \le 1, \quad t > 0, \tag{2}$$

and the boundary conditions

$$h(0,t) = h_x(0,t) = 0 \qquad \text{and} \qquad M(1,t) = Q(1,t) = 0, \tag{3}$$

where $M(x,t)$ and $Q(x,t)$ are the moment and the shear, respectively [10]:

$$M(x,t) = E(x)I(x)h_{xx}(x,t) \qquad \text{and} \qquad Q(x,t) = M_x(x,t). \tag{4}$$

Now we replace the free right-end conditions from Eq. (3) with the following boundary feedback control law [2, 4]. Define the input and the output as

$$U(t) = [-Q(1,t),\ M(1,t)]^T \qquad \text{and} \qquad Y(t) = [h_t(1,t),\ h_{xt}(1,t)]^T, \tag{5}$$

where T stands for transposition. The feedback control law is given by

$$U(t) = KY(t), \tag{6}$$

where K is the 2×2 feedback matrix. We select

$$K = \text{codiag}(-k_2, -k_1), \qquad k_1, k_2 \ge 0, \tag{7}$$

with k_1,k_2 being the control parameters. The feedback (6) can be written as

$$E(1)I(1)h_{xx}(1,t) = -k_1 h_t(1,t) \quad \text{and} \quad (E(x)I(x)h_{xx}(x,t))_x|_{x=1} = k_2 h_{xt}(1,t). \tag{8}$$

Finally, we arrive to the following initial-boundary value problem: the equation of motion (2), the boundary conditions (3), and the standard initial conditions $h(x,0) = h_0(x)$, $h_t(x,0) = h_1(x)$.

Notice that the choice of a feedback matrix K defines whether the system is dissipative or not. Indeed, let $\mathcal{E}(t)$ be the energy of the system, defined by representation (1). Evaluating $\mathcal{E}_t(t)$ on the solutions of Eq. (2) satisfying the left-end conditions from Eqs. (3), we obtain

$$\begin{aligned}\mathcal{E}_t(t) &= \int_0^1 [\varrho(x)A(x)h_t(x,t)h_{tt}(x,t) + E(x)I(x)h_{xx}(x,t)h_{xxt}(x,t)]x \\ &= -(E(x)I(x)h_{xx}(xt))_x h_t(xt)|_{x=1} + E(1)I(1)h_{xx}(1,t)h_{xt}(1,t).\end{aligned} \tag{9}$$

Taking into account Eqs. (4) and (6), we represent the right-hand side of Eq. (9) as the dot product in $\mathbb{R}^2$:

$$\mathcal{E}_t(t) = -Q(1,t)h_t(1,t) + M(1,t)h_{xt}(1,t) = U(t)\cdot Y(t) = KY(t)\cdot Y(t). \tag{10}$$

With the choice of K as in Eq. (7), we have

$$\mathcal{E}_t(t) = \begin{bmatrix} 0 & -k_2 \\ -k_1 & 0 \end{bmatrix}\begin{bmatrix} 2h_t(1,t) \\ h_{xt}(1,t)\end{bmatrix}\cdot\begin{bmatrix} 2h_t(1,t) \\ h_{xt}(1,t)\end{bmatrix} = -(k_1+k_2)h_t(1,t)h_{xt}(1,t). \tag{11}$$

Thus the system is not dissipative for all nonnegative values of k_1 and k_2.

2. OPERATOR FORM OF THE PROBLEM

In what follows, we incorporate the cross-sectional area $A(x)$ into the density, write $\rho(x)$ instead of $Q(x)A(x)$, and also abbreviate $EI(x) \equiv E(x)I(x)$. Let $\mathcal{H}$ be the Hilbert space of two-component vector functions $U(x) = [u_0(x), u_1(x)]^T$ equipped with the following norm:

$$\|U\|^2_{\mathcal{H}} = \frac{1}{2}\int_0^1 \left[EI(x)|u_0''(x)|^2 + \rho(x)|u_1(x)|^2\right]x. \tag{12}$$

Assuming that $EI, \rho \in C^2(0,1)$ are positive functions, we obtain that the closure of smooth functions $U(x) = [u_0(x), u_1(x)]^T$ satisfying $u_0(0) = u_0'(0) = 0$ will produce the energy space $\mathcal{H} = H_0^2(0,1) \times L^2(0,1)$. Here $H_0^2(0,1) = \{u \in H^2(0,1) : u(0) = u'(0) = 0\}$, and the equality of function spaces is understood in the sense of a Hilbert-space isomorphism.

Problem (2) with conditions (3) can be represented as the time evolution problem:

$$U_t(x,t) = i(\mathcal{L}_{k_1,k_2}U)(x,t) \qquad \text{and} \qquad U(x,0) = [u_0(x), u_1(x)]^T, \tag{13}$$

where $0 \geq x \geq 1,\ t \geq 0$. The dynamics generator $\mathcal{L}_{k1,k2}$ is given by the following matrix differential expression:

$$\mathcal{L}_{k_1,k_2} = -i\begin{bmatrix} 0 & 1 \\ -\dfrac{1}{\rho(x)}\dfrac{\partial^2}{\partial x^2}\left(EI(x)\dfrac{\partial^2}{\partial x^2}\right) & 0\end{bmatrix}, \tag{14}$$

Defined on the domain

$$\mathcal{D}(\mathcal{L}_{k_1,k_2}) = \{U = (u_0, u_1)^T \in \mathcal{H} : u_0 \in H^4(0,1),\ \ u_1 \in H_0^2(0,1);\ \ u_1(0) = u_1'(0) = 0;$$

$$EI(1)u_0''(1) = -k_1 u_1(1),\ \ (EI(x)u_0''(x))'\big|_{x=1} = k_2 u_1'(1)\}. \quad (15)$$

For any $(k_1,k_2) \in \mathbb{R}^2$, the adjoint operator $\mathcal{L}^*_{k_1,k_2}$ [13] is given by

$$\mathcal{L}^*_{k_1,k_2} = \mathcal{L}_{-k_2,-k_1}, \quad (16)$$

i.e., $\mathcal{L}^*_{k1,k2}$ is defined by the same differential expression (14) on the domain described in Eq. (15), where k_1 and k_2 are replaced by $(-k_2)$ and $(-k_1)$, respectively. It follows from Eq. (16) that $\mathcal{L}_{0,0}$ is self-adjoint in $\mathcal{H}$ and thus $\mathcal{L}_{0,0}$ is the dynamics generator of the clamped-free beam model. For the reader's convenience, we summarize the properties of $\mathcal{L}_{k1,k2}$ from [1, 2] needed for the present work.

Proposition 1:

1. $\mathcal{L}_{k1,k2}$ is an unbounded operator with compact resolvent, whose spectrum consists of a countable set of normal eigenvalues (i.e., isolated eigenvalues, each of finite algebraic multiplicity [6, 13]).
2. For each $(k_1, k_2) \in \mathbb{R}^2,\ |k_1| + |k_2| > 0,$, the operator $\mathcal{L}_{k1,k2}$ is a rank-two perturbation of the self-adjoint operator $\mathcal{L}_{0,0}$ in the sense that the operators $\mathcal{L}^{-1}_{k1,k2}$ and $\mathcal{L}^{-1}_{0.0}$ exist and are related by the rule

$$\mathcal{L}^{-1}_{k_1,k_2} = \mathcal{L}^{-1}_{0,0} + \mathcal{T}_{k_1,k_2}, \quad (17)$$

where $\mathcal{T}_{k1,k2}$ is a rank-two operator. A similar decomposition is valid for the adjoint operator, i.e.,

$$\left(\mathcal{L}^{-1}_{k_1,k_2}\right)^* = \mathcal{L}^{-1}_{0,0} + \mathcal{T}^*_{k_1,k_2}, \qquad \mathcal{T}^*_{k_1,k_2} = \mathcal{T}_{-k_2,-k_1}. \quad (18)$$

From now on, we assume that the structural parameters are constant. In the case of variable parameters, the spectral asymptotics will have the same leading terms and remainder terms depending on parameter smoothness.

Proposition 2: Assume that k_1, k_2>0 and k_1k_2≠EIρ. Let

$$\mathcal{K} = \frac{k_1 + k_2}{A - k_1k_2/A}, \qquad A = \sqrt{EI\rho}, \qquad \text{and} \qquad |\mathcal{K}| \neq 1. \quad (19)$$

The following asymptotic approximations for the eigenvalues λ_n (as $|n|\to\infty$) of the operator $\mathcal{L}_{k1,k2}$ hold:

1. If $1<|\mathcal{K}|<\infty$, then for $|n|\to\infty$ one has

$$\lambda_n = \operatorname{sign}(\mathcal{K}n)\sqrt{\frac{EI}{\rho}}\left[(\pi n)^2 - \frac{1}{4}\ln^2\left(\frac{\mathcal{K}+1}{\mathcal{K}-1}\right) + i\pi n\ln\left(\frac{\mathcal{K}+1}{\mathcal{K}-1}\right)\right] + O(ne^{-\pi|n|}). \quad (20)$$

2. If $0<\mathcal{K}<1$, then for $n\to\infty$ one has

$$\lambda_n = \text{sign}(\mathcal{K}n)\sqrt{\frac{EI}{\rho}}\left[\left(\tfrac{2n+1}{2}\pi\right)^2 - \frac{1}{4}\ln^2\left(\frac{\mathcal{K}+1}{\mathcal{K}-1}\right) + i\pi\left(\frac{2n+1}{2}\right)\ln\left(\frac{\mathcal{K}+1}{\mathcal{K}-1}\right)\right] + O(ne^{-\pi|n|}) \tag{21}$$

First of all, we address the question of accuracy of the asymptotic formulas (20) and (21). By its nature, formula (20) (as well as formula (21)) means that for any small ε>0, one can find a positive integer *N*, such that all eigenvalues λ_n with $|n|\geq N+1$ satisfy the estimate

$$\left|\lambda_n - \text{sign}(\mathcal{K}n)\sqrt{\frac{EI}{\rho}}\left[(\pi n)^2 - \frac{1}{4}\ln^2\left(\frac{\mathcal{K}+1}{\mathcal{K}-1}\right) + i\pi n\ln\left(\frac{\mathcal{K}+1}{\mathcal{K}-1}\right)\right]\right| \leq \varepsilon \tag{22}$$

for the case when $1<|\mathcal{K}|<\infty$ and

$$\left|\lambda_n - \text{sign}(\mathcal{K}n)\sqrt{\frac{EI}{\rho}}\left[\left(\frac{2n+1}{2}\pi\right)^2 - \frac{1}{4}\ln^2\left(\frac{\mathcal{K}+1}{\mathcal{K}-1}\right) + i\pi\left(\frac{2n+1}{2}\right)\ln\left(\frac{\mathcal{K}+1}{\mathcal{K}-1}\right)\right]\right| \leq \varepsilon \tag{23}$$

for the case when $0<|K|<1$. The following important question holds: *From which index N can the eigenvalues be approximated by the leading asymptotic terms with acceptable accuracy?* In other words, can one claim that the asymptotic formulas (20) and (21) are valuable to practitioners, or are they just important mathematical results of the spectral analysis?

The results of numerical simulations (see Tables 1 and 2 and Figures 1 and 2) show that the asymptotic formulas are indeed quite accurate. That is, if one places on the complex plane the numerically produced sets of the eigenvalues, then the theoretically predicted distribution of eigenvalues can be seen almost immediately. To obtain these results, we used the numerical procedure based on Chebyshev polynomial approximations [14, 15, 16], as outlined in Section 5.

In Figures 1 and 2, we represent the graphical distribution of the eigenvalues corresponding to the discretized operator ("numerical" eigenvalues) and the leading asymptotic terms from Eqs. (20) and (21) ("analytic" eigenvalues). In Tables 1 and 2, the numerical values of the corresponding graphical points on Figures 1 and 2 are listed. We have used the following notations: *N*=64 is the number of grid points on [0,1], and ε*f* is the filtering parameter as described in Eq. (69). It can be easily seen from the graphs and tables that the two sets of data coincide almost immediately, i.e., the leading asymptotic terms in the approximations are very close to the numerically approximated eigenvalues.

Figure 3 shows the sub-domains of the k_1,k_2-plane, which correspond to different intervals for the values of $\mathcal{K}$ defined by Eq. (19). On the sub-domain with dark gray color $\mathcal{K}$ such that $|\mathcal{K}|>1$, i.e., to evaluate the asymptotic approximation for the eigenvalues, one needs formula (20), while on the complementary sub-domain, one needs formula (21).

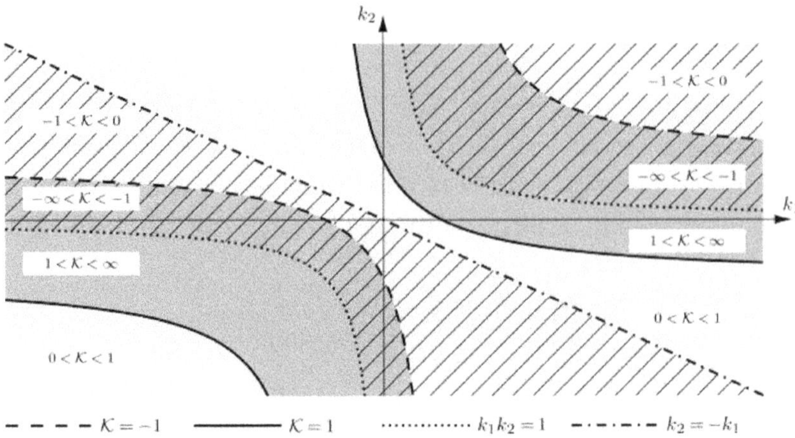

Figure 3. Regions of $\mathcal{K}$ on the k_1,k_2-plane, $A = \sqrt{EI\rho} = 1$.

3. THE DEADBEAT MODES

An eigenvalue λ_n of the dynamics generator $\mathcal{L}_{k1,k2}$ is called a deadbeat mode if $\lambda_n = i\beta_n, \beta_n > 0$. If the corresponding eigenfunction is $\Phi_n(x)$, then the evolution problem (13) has a solution given in the form $e^{i\lambda_n t}\Phi_n(x) = e^{-\beta_n t}\Phi_n(x)$, which tends to zero without any oscillation.

As shown in paper [2], for the case when one of the control parameters is zero and the other one is positive, the entire set of the eigenvalues is located in the closed upper half plane. This result is not obvious since the operator is not dissipative; in fact, it requires a fairly nontrivial proof. However, due to this fact, we assume that any deadbeat mode can be given in the form $i\beta$, with $\beta > 0$. To deal with the deadbeat modes analytically, we rewrite the spectral equation $(\mathcal{L}_{k_1,k_2}\Phi)(x) = \lambda\Phi(x)$ in the form of an equivalent problem for an operator pencil [17] as

$$\begin{aligned} EI\varphi''''(x) &= \lambda^2\rho\varphi(x), & \varphi(0) &= \varphi'(0) = 0, \\ EI\varphi''(1) &= -i\lambda k_1\varphi(1), & EI\varphi'''(1) &= i\lambda k_2\varphi'(1). \end{aligned} \tag{24}$$

If λ_n and $\varphi_n(x)$ are an eigenvalue and eigenfunction of the pencil (24), then λ_n is also an eigenvalue of $\mathcal{L}_{k1,k2}$ with the eigenfunction $\Phi_n(x) = \left[\frac{1}{i\lambda_n}\varphi_n(x), \varphi_n(x)\right]^T$.

To solve problem (24), we first redefine the spectral and control parameters to eliminate ρ and EIEI from Eq. (24). We define $\tilde{\lambda}$, $\tilde{k}_1$, and $\tilde{k}_2$ by $\lambda = \sqrt{EI/\rho}\tilde{\lambda}$ and $\tilde{k}_j = \sqrt{EI\rho}k_j, j = 1, 2$. Substituting these relations into Eq. (24) and eliminating the "tilde," we obtain the following Sturm-Liouville eigenvalue problem:

$$\varphi''''(x) = \lambda^2\varphi(x), \quad \varphi(0) = \varphi'(0) = 0, \quad \varphi''(1) = -i\lambda k_1\varphi(1), \quad \varphi'''(1) = i\lambda k_2\varphi'(1). \tag{25}$$

The solution of Eq. (25) satisfying the left-end boundary conditions $\varphi(0) = \varphi'(0) = 0$ can be written in the form

$$\varphi(\lambda,x) = \mathcal{A}(\lambda)\left[\cosh\left(\sqrt{\lambda}x\right) - \cos\left(\sqrt{\lambda}x\right)\right] + \mathcal{B}(\lambda)\left[\sinh\left(\sqrt{\lambda}x\right) - \sin\left(\sqrt{\lambda}x\right)\right]. \tag{26}$$

Substituting formula (26) into the right-end boundary conditions of Eq. (25), one gets a system for the coefficients $\mathcal{A}(\lambda)$ and $\mathcal{B}(\lambda)$:

$$\begin{aligned} &\mathcal{A}(\lambda)\left[(1+ik_1)\cosh\sqrt{\lambda} - (1-ik_1)\cos\sqrt{\lambda}\right] + \\ &\quad \mathcal{B}(\lambda)\left[(1+ik_1)\sinh\sqrt{\lambda} - (1-ik_1)\sin\sqrt{\lambda}\right] = 0, \\ &\mathcal{A}(\lambda)\left[(1-ik_2)\sinh\sqrt{\lambda} - (1+ik_2)\sin\sqrt{\lambda}\right] + \\ &\quad \mathcal{B}(\lambda)\left[(1-ik_2)\cosh\sqrt{\lambda} + (1+ik_2)\cos\sqrt{\lambda}\right] = 0. \end{aligned} \tag{27}$$

Let $\Delta(\lambda)$ be the determinant of the matrix of coefficients for $\mathcal{A}(\lambda)$ and $\mathcal{B}(\lambda)$ in Eqs. (27). System (27) has nontrivial solutions if and only if $\Delta(\lambda)=0$, i.e.,

$$(1+k_1k_2) + (1-k_1k_2)\cosh\sqrt{\lambda}\cos\sqrt{\lambda} + i(k_1+k_2)\sinh\sqrt{\lambda}\sin\sqrt{\lambda} = 0. \tag{28}$$

Theorem 1: The following results hold in the case when $k_1>0$ and $k_2=0$. Similar results hold in the case when $k_1=0$ and $k_2>0$.

1. For $0<k_1<1$, the deadbeat modes do not exist.
2. For $k_1=1$, there exist infinitely many deadbeat modes given explicitly by
$$\lambda_n = \mu_n^2, \qquad \mu_n = x_n(1+i), \qquad x_n = \frac{\pi}{2}(2n+1), \qquad n = 0,1,2,\dots \tag{29}$$
3. For any $k_1>1$, there exist a finite number $N(k_1)$ of deadbeat modes. Each mode has the form $\lambda=\mu^2, \mu=x(1+i)$, where x is a root of the function
$$H(x;k_1) \equiv 2 + (1+k_1)\cos 2x + (1-k_1)\cosh 2x, \quad x>0. \tag{30}$$

Let $X(k_1) = \frac{1}{2\pi}\cosh^{-1}\left(1+\frac{4}{k_1-1}\right)$, and then $N(k_1)$ satisfies the estimate

$$2[X(k_1)] + 1 \le \mathcal{N}(k_1) \le 2[X(k_1)] + 3. \tag{31}$$

Hence $\mathcal{N}(k_1)\to\infty$ as $k_1\to 1^+$. (By $[X]$ we denote the greatest integer less than or equal to X).

Proof: Let $\mu = \sqrt{\lambda} = x(1+i), x>0.$. Taking into account the relations

$$\begin{aligned} 2\cosh\mu\cos\mu &= \cosh(1+i)\mu + \cosh(1-i)\mu, \\ 2i\sinh\mu\sin\mu &= \cosh(1+i)\mu - \cosh(1-i)\mu, \end{aligned}$$

we reduce Eq. (28) to the following form:

$$2 + (1+k_1)\cos 2x + (1-k_1)\cosh 2x = 0, \quad x>0. \tag{32}$$

It can be readily seen that if $0<k_1<1$, then $2+(1+k_1)\cos 2x>0$ and $(1-k_1)\cosh 2x>0$, which means that Eq. (32) has no solutions. Statement (1) is shown. Statement (2) follows immediately if one considers Eq. (32) for $k_1=1$.

To prove Statement (3), we rewrite Eq. (32) in the form

$$\cosh 2x = \frac{2}{k_1 - 1} + \left(1 + \frac{2}{k_1 - 1}\right)\cos 2x, \quad x > 0, \quad k_1 > 1. \tag{33}$$

The left-hand side of Eq. (33) is monotonically increasing, while the right-hand side is sinusoidal, with maximum $\left(1+\frac{4}{k_1-1}\right)$ and minimum (-1), and period π. So the graphs of the left- and right-hand side have intersections only on the interval $\left[0, \frac{1}{2}\cosh^{-1}\left(1+\frac{4}{k_1-1}\right)\right]$. There are two intersections for each full period of the right-hand side that fits into the above interval (Figure 4). As it can be seen in Figure 4, one should add at least one more intersection for the first half-period after the full periods. Depending on the value of k_1, the two graphs can have two intersections, one tangential intersection or no intersections on the second half-period. This leads to estimate (31).

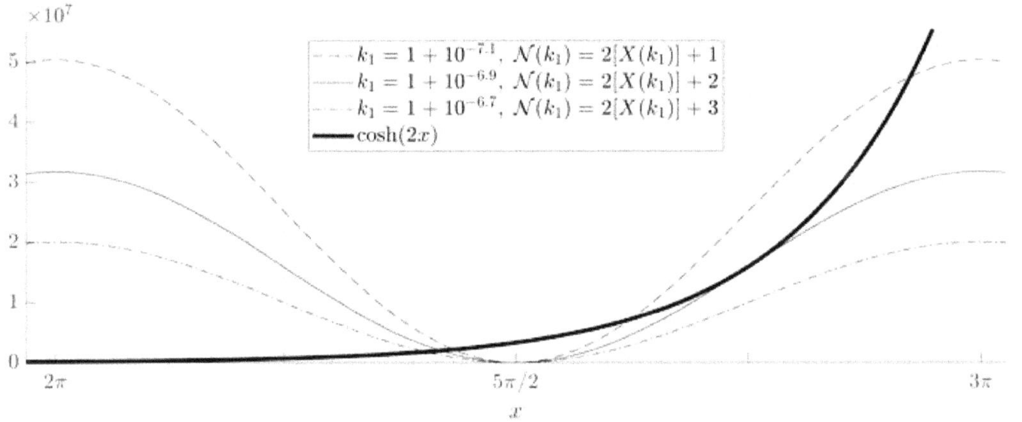

Figure 4. Left- and right-hand side of Eq. (33) for different values of k_1.

A graphical illustration of the result of Theorem 1 is shown in Figure 5.

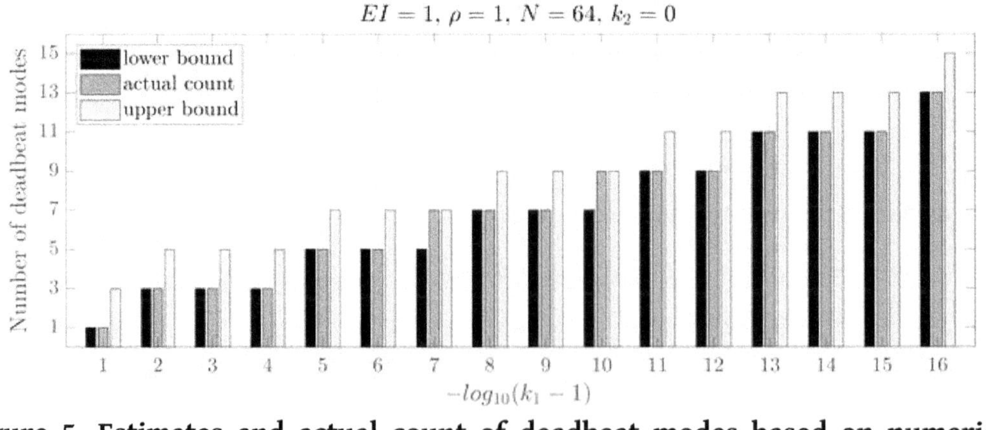

Figure 5. Estimates and actual count of deadbeat modes based on numerical simulations.

4. STRUCTURE OF THE DEADBEAT MODE SET

The main result on the existence and distribution of double roots of the function $H(x; k_1)$ is presented in the statement below.

Theorem 2: For a given $k=>1$, the multiplicity of each root of $H(x;k_1)$ does not exceed 22. There exists a sequence $\left\{k_1^{(n)}; n = 0, 1, 2, \dots\right\}$, such that the function $H(x;k_1)$ has a double root if and only if $k_1 = k_1^{(n)}$ for some n. So the original spectral problem with $k_1 = k_1^{(n)}, k_2 = 0$ has a double deadbeat mode $\lambda_n = \mu_n^2 = 2ix_n^2$. The following asymptotic formulas hold

$$x_n = \frac{3\pi}{4} + \pi n + P_n^{-1} + O(P_n^{-2}), \qquad \text{where} \qquad P_n = \exp\left\{\frac{3\pi}{2} + 2\pi n\right\}, \qquad (34)$$

and

$$k_1^{(n)} = 1 + 4P_n^{-1} + O(P_n^{-2}). \qquad (35)$$

Proof: If x is a double root of H, then $H(x;k_1)=H'(x;k_1)=0$, i.e., separating the real and imaginary parts, we have

$$2 + (k_1 + 1)\cos 2x - (k_1 - 1)\cosh 2x = 0, \qquad (36)$$

$$(k_1 + 1)\sin 2x + (k_1 - 1)\sinh 2x = 0. \qquad (37)$$

Eliminating k_1 from system given by (36) and (37), we obtain that the following equation has to be satisfied:

$$G(x) \equiv (1 + \cos 2x)\sinh 2x + (1 + \cosh 2x)\sin 2x = 0, \qquad x > 0. \qquad (38)$$

Rewriting Eq. (37) in the form $(k_1+1)\sin 2x=-(k_1-1)\sinh 2x$, and taking into account that $k_1>1$, we obtain that if x is the solution of Eq. (38), then $\sin 2x<0$.

Now we show that when $\cos 2x<0$ and $\sin 2x<0$, i.e.,

$$\pi(2n + 1) < 2x < \frac{3\pi}{2} + 2\pi n, \qquad n \in \{0, 1, 2, \dots\},$$

Eq. (38) does not have any solutions. Indeed, in the above range of x, we have $\cos 2x + \sin 2x = \sqrt{2}\sin(2x + \pi/4) < -1$ and $|\sin 2x-\cos 2x|<1$. With such estimates we obtain that

$$G(x) = \sin 2x + \sinh 2x + \frac{1}{2}e^{2x}(\cos 2x + \sin 2x) + \frac{1}{2}e^{-2x}(\sin 2x - \cos 2x) < \sin 2x < 0, \qquad (39)$$

which mean that Eq. (38) cannot be satisfied.

Now we consider the case when $\cos 2x>0$ and $\sin 2x<0$, i.e.,

$$\frac{3\pi}{2} + 2\pi n < 2x < 2\pi(n + 1), \qquad n \in \{0, 1, 2, \dots\}.$$

It is convenient to rewrite system given by (36) and (37) in the form $2x=3\pi/2+2\pi n+s$, where $n\in\{0,1,2,...\}$, $0<s<\pi/2$. If $g(s)\equiv G(3\pi/4+\pi n+s)$, then Eq. (38) generates the following equation for g:

$$g(s) \equiv (1+\sin s)\sinh\left(\frac{3\pi}{2}+2\pi n+s\right)-\left[1+\cosh\left(\frac{3\pi}{2}+2\pi n+s\right)\right]\cos s=0. \tag{40}$$

Let us show that for each n, Eq. (40) has a unique solution. For $s=0$ we have

$$g(0)=\sinh\left(\frac{3\pi}{2}+2\pi n\right)-1-\cosh\left(\frac{3\pi}{2}+2\pi n\right)<0,$$

and for $s=\pi/2$ we have $g(\pi/2)=2\sinh(2\pi(n+1))>0$. Evaluating g' we have

$$g'(s)=\sin s+(1+2\sin s)\cosh\left(\frac{3\pi}{2}+2\pi n+s\right)>0. \tag{41}$$

Thus $g(s)$ is a monotonically increasing function, such that $g(0)<0<g(\pi/2)$, which means that g has a unique root on $[0,\pi/2]$.

Finally we show that the multiplicity of a multiple root cannot exceed 2. Using a contradiction argument, assume that there exists x_0, such that in addition to Eqs. (36) and (37), one has $H''(x_0;k_1)=0$, i.e., the multiplicity of x_0 is at least 3. The system $H''(x_0;k_1)=0$and $H''(x_0;k_1)=0$ can be written as

$$\begin{aligned}(k_1+1)\sin 2x_0+(k_1-1)\sinh 2x_0 &= 0,\\ (k_1+1)\cos 2x_0+(k_1-1)\cosh 2x_0 &= 0.\end{aligned} \tag{42}$$

Since $k_1>1$, the second equation of (42) yields $\cos 2x_0<0$. Also, since x_0 is a multiple root, we must have $\sin 2x_0<0$. Then $2x_0$ is in the third quadrant, which means that $G(x_0)\neq 0$, as we have seen above. This contradicts our assumption that x_0 is a root of Eqs. (36) and (37).

To derive asymptotic distribution of the roots of Eq. (40), we check that with P_n from Eq. (34), the following approximations are valid:

$$\begin{aligned}&\sin(2P_n^{-1})=2P_n^{-1}+O(P_n^{-3}),\quad \cos(2P_n^{-1})=1-2P_n^{-2}+O(P_n^{-4}),\\ &2\sinh\left(\frac{3\pi}{2}+2\pi n+2P_n^{-1}\right)=P_n+2+P_n^{-1}+O(P_n^{-2}),\\ &2\cosh\left(\frac{3\pi}{2}+2\pi n+2P_n^{-1}\right)=P_n+2+3P_n^{-1}+O(P_n^{-2}).\end{aligned} \tag{43}$$

Evaluating $g(s)$ from Eq. (40) for $s=2^{P-1}{}_n$ and using Eq. (43), we get

$$\begin{aligned}g(2P_n^{-1})=&\left[1+\sin(2P_n^{-1})\right]\sinh\left(\frac{3\pi}{2}+2\pi n+2P_n^{-1}\right)-\\ &\left[1+\cosh\left(\frac{3\pi}{2}+2\pi n+2P_n^{-1}\right)\right]\cos(2P_n^{-1})=2P_n^{-1}+O(P_n^{-2}).\end{aligned} \tag{44}$$

Representation (44) implies that there exists n_0, such that for all $n \geq n_0$, we have $g(2P_n^{-1}) > 0$. Taking into account that $g(0)<0$, we obtain that the root s_n, $n \geq n_0$, of the function $g(s)$ is located on the interval $(0, 2P_n^{-1})$. To find the location of this root more precisely [18], we use linear interpolation. Namely, substituting Eq. (43) into the expression for $g'(s)$ from Eq. (41) yields

$$g'(2P_n^{-1}) = \frac{P_n}{2} + O(1). \tag{45}$$

Replacing $g(s)$ by the linear function tangential to $g(s)$ at the point $(2P_n^{-1}, g(2P_n^{-1}))$, and finding the root of this function, we get

$$s_n = 2P_n^{-1} - \frac{g(2P_n^{-1})}{g'(2P_n^{-1})} + O(P_n^{-2}) = 2P_n^{-1} + O(P_n^{-2}). \tag{46}$$

Having this approximation for s_n, we immediately get

$$x_n = \frac{3\pi}{4} + \pi n + \frac{s_n}{2} = \frac{3\pi}{4} + \pi n + P_n^{-1} + O(P_n^{-2}). \tag{47}$$

From the equation $H\left(x_n; k_1^{(n)}\right) = 0$, we obtain the formula for $k_1^{(n)}$ as

$$k_1^{(n)} = \frac{\sinh(3\pi/2 + 2\pi n + s_n) + \cos s_n}{\sinh(3\pi/2 + 2\pi n + s_n) - \cos s_n}. \tag{48}$$

Substituting formulas (43) and (46) into formula (48), we obtain representation (35).

Corollary 1: Let $k_1 = k_1^{(n)}$ for some $n \in \mathbb{N}^+ \cup \{0\}$, and let x_n be the corresponding double root of the function $H\left(x; k_1^{(n)}\right)$. Then $\lambda_0 = 2ix_n^2$ is an eigenvalue of the operator $\mathcal{L}_{k_1^{(n)}, 0}$, such that the geometric multiplicity of λ_0 is 1 and its algebraic multiplicity is 2. Therefore there exists a unique eigenvector Φ and one associate vector Ψ, such that

$$\mathcal{L}_{k_1^{(n)}, 0}\Phi = \lambda_0 \Phi, \qquad \mathcal{L}_{k_1^{(n)}, 0}\Psi - \lambda_0 \Psi = \Phi. \tag{49}$$

Proof: It suffices to show that problem (24) does not have two linearly independent eigenvectors corresponding to λ_0 [18, 19]. Using contradiction argument we assume that for some λ_0 the boundary-value problem (25) with $k_2=0$ has two linearly independent solutions ψ and χ. Each function satisfies the problem

$$\varphi''''(x) = \lambda_0^2 \varphi(x), \qquad \varphi(0) = \varphi'(0) = 0, \qquad \varphi'''(1) = 0, \qquad \varphi''(1) = i\lambda_0 k_1 \varphi(1). \tag{50}$$

First we observe that $\psi(1)\chi(1) \neq 0$. Indeed, if $\psi(1)=0$, then we have

$$\int_0^1 |\psi''(\sigma)|^2 d\sigma = \int_0^1 \psi''''(\sigma)\overline{\psi(\sigma)} d\sigma = \lambda_0^2 \int_0^1 |\psi(\sigma)|^2 d\sigma. \tag{51}$$

Since λ_0 is purely imaginary, Eq. (51) is not valid. We define a new function:

$$g(x) = \psi(x) - \frac{\psi(1)}{\chi(1)}\chi(x). \tag{52}$$

One can readily check that g satisfies the following boundary-value problem:

$$\varphi''''(x) = \lambda_0^2\varphi(x), \qquad \varphi(0) = \varphi'(0) = 0, \qquad \varphi(1) = \varphi''(1) = 0, \tag{53}$$

and therefore

$$\int_0^1 |g''(\sigma)|^2 d\sigma = \int_0^1 g''''(\sigma)\overline{g(\sigma)}\sigma = \lambda_0^2 \int_0^1 |g(\sigma)|^2 d\sigma. \tag{54}$$

Eq. (54) is valid if and only if $\lambda_0^2 > 0$; however, for a deadbeat mode, $\lambda_0^2 > 0$; The obtained contradiction means that for each double mode, there is one eigenfunction and one associate function.

4.1 Deadbeat mode behavior as $k_1 \to 1$

As k_1 approaches 11, the spectral branches are moving upward and toward the imaginary axis (Figures 6 and 7). As a result of this motion, eigenvalues approach the imaginary axis at different rates depending on whether k_1 approaches 11 from above or below.

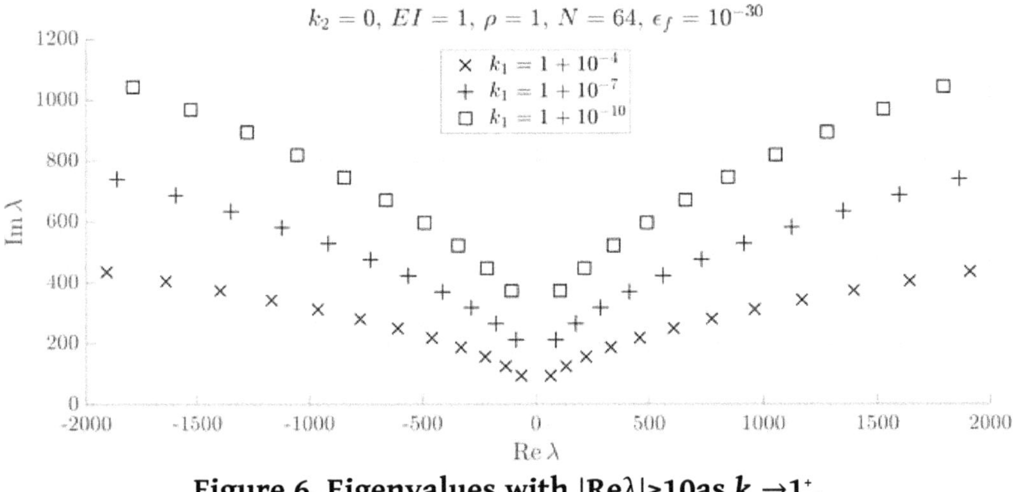

Figure 6. Eigenvalues with |Reλ|>10as $k_1 \to 1^+$.

As follows from Table 3, the real parts of the eigenvalues decrease steadily as $k_1 \to 1^+$, to a point where the eigenvalue becomes a deadbeat mode. An increase in the number of deadbeat modes can be seen as $k_1 \to 1^+$, which is in agreement with Statement (3) of Theorem 1. One can see from Table 3 that there are pairs of modes such that the distance between them tends to zero as $k_1 \to 1^+$. (Compare modes no.5 and no.7 for $|k_1-1|=10^{-4}$, modes no.4 and no.7 for $|k_1-1|=10^{-6}$, modes no.5 and no.8 for $|k^1-1|=10^{-8}$, and modes no.4 and no.7 for $|k_1-1|=10^{-10}$. Such behavior indicates convergence of the two simple deadbeat modes to a double mode, which is consistent with Theorem 2.

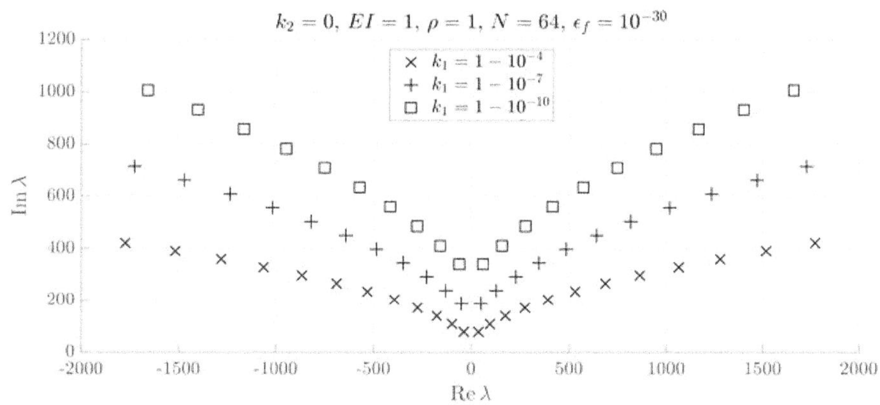

Figure 7. Eigenvalues with |Reλ|>10as $k_1 \to 1^-$.

Analyzing Table 4, one can see that the eigenvalues get closer to the imaginary axis as $k_1 \to 1^-$. However the rate at which their real parts approach zero is significantly lower than in the case $k_1 \to 1^+$. Even at $k_1=1-10^{-10}$, the eigenvalue closest to the imaginary axis has a real part of about 10^{-4}, which means that it is not a deadbeat mode (see Statement (1) of Theorem 1).

The eigenvalues near the imaginary axis approach the same double deadbeat modes in both cases when $k_1 \to 1^\pm$ (see Statement (2) of Theorem 1). In conclusion, one can claim that the eigenvalues are indeed approaching the imaginary axis; however, the rate of this approach is different for $k_1 \to 1^-$ and $k_1 \to 1^+$. In the former case, an eigenvalue's distance from the imaginary axis decreases very slowly; in the latter case, the eigenvalues quickly "jump" on the imaginary axis and turn into deadbeat modes.

5. OUTLINE OF THE NUMERICAL SCHEME

To carry out the numerical analysis of the differential operator $\mathcal{L}_{k1,k2}$, we use the Chebyshev collocation method and cardinal functions [14, 15, 16].

Recall that the Nth Chebyshev polynomial of the first kind is defined by

$$T_N(\xi) = \cos N\theta, \quad -1 \le \xi \le 1 \quad \text{where} \quad \xi = \cos\theta, \quad \theta \in [0, \pi]. \tag{55}$$

The cardinal functions, $\psi_k(\xi)$, and the Chebyshev-Gauss-Lobatto (CGL) grid points $\{\xi_k\}$ are defined as follows:

$$\psi_k(\xi) = (-1)^k \frac{(1-\xi^2)T'_{N-1}(\xi)}{c_k(N-1)^2(\xi-\xi_k)}, \qquad \xi_k = \cos\frac{(k-1)\pi}{N-1}, \quad \text{for} \quad 1 \le k \le N, \tag{56}$$

where coefficients c_k are such that $c_1=c_N=2$ and $c_k=1$ for $1<k<N$. The main property of cardinal functions is $\psi_k(\xi_j) = \delta_{kj}$ (using the Kronecker delta). The family $\{\psi_k\}_{k=1}^N$ forms a basis in the space of polynomials of degree $(N-1)$, i.e., if f is such polynomial, then f and f' can be written in the forms

$$f(\xi) = \sum_{k=1}^{N} f(\xi_k)\psi_k(\xi) \qquad \text{and} \qquad f'(\xi) = \sum_{k=1}^{N} f'(\xi_k)\psi_k(\xi). \tag{57}$$

If $f = [f(\xi_1),\ f(\xi_2),\ ...,\ f(\xi_N)]^T$ and $g = [f'(\xi_1),\ f'(\xi_2),\ ...,\ f'(\xi_N)]^T$, then g=Df, where D is the Chebyshev derivative matrix with the elements

$$D_{11} = -D_{NN} = \frac{1+2(N-1)^2}{6}, \qquad D_{kk} = -\frac{\xi_k}{2(1-\xi_k^2)} \quad \text{for} \quad 1<k<N,$$
$$D_{j,k} = \frac{c_j(-1)^{j+k}}{c_k\left(\xi_j - \xi_k\right)} \quad \text{for} \quad j \neq k. \tag{58}$$

5.1 Discretization of $\mathcal{L}_{k1,k2}$

Rescaling the independent variable x as ξ=2x−1, we rewrite the operator and its domain, representations (14) and (15), in the form

$$\mathcal{L}_{k_1,k_2} = -i\begin{bmatrix} 0 & 1 \\ -\dfrac{16}{\rho(\xi)}\dfrac{\partial^2}{\partial\xi^2}\left(EI(\xi)\dfrac{\partial^2}{\partial\xi^2}\right) & 0 \end{bmatrix}, \tag{59}$$

and

$$\mathcal{D}(\mathcal{L}_{k_1,k_2}) = \{(u_0,u_1)^T \in \mathcal{H}:\ u_0 \in H^4(-1,1),\ u_1 \in H_0^2(-1,1);\ u_1(-1) = u_1'(-1) = 0;$$
$$4EI(1)u_0''(1) = -k_1u_1(1),\ 4\left(EI(\xi)u_0''(\xi)\right)'\Big|_{\xi=1} = k_2u_1'(1)\}, \tag{60}$$

where $\mathcal{H} = H_0^2(-1,1)\times L^2(-1,1)$, equipped with the norm

$$\|U\|_{\mathcal{H}}^2 = \frac{1}{4}\int_{-1}^{1}\left[16EI(\xi)\left|u_0''(\xi)\right|^2 + \rho(\xi)\left|u_1(\xi)\right|^2\right]d\xi. \tag{61}$$

We approximate the action of $\mathcal{L}_{k1,k2}$ on the finite-dimensional subspace $\mathcal{H}_N \subset \mathcal{H}$ of polynomials of degree at most (N-1). Using the CGL grid and the cardinal functions, we substitute for u_0 and u_1 their truncated expansions:

$$u_0(\xi) \approx \sum_{k=1}^{N}\Phi_k\psi_k(\xi), \quad \Phi_k = u_0(\xi_k), \qquad u_1(\xi) \approx \sum_{k=1}^{N}\Theta_k\psi_k(\xi), \quad \Theta_k = u_1(\xi_k). \tag{62}$$

Let Φ and Θ be N-dim vectors and Ψ be a 2N-dim vector defined by

$$\mathbf{\Phi} = [\Phi_1,\ \Phi_2,\ ...,\ \Phi_N]^T, \qquad \mathbf{\Theta} = [\Theta_1,\ \Theta_2,\ ...,\ \Theta_N]^T, \qquad \mathbf{\Psi} = \begin{bmatrix}\mathbf{\Phi}\\ \mathbf{\Theta}\end{bmatrix}. \tag{63}$$

Let L be the finite-dimensional approximation of the differential operator $\mathcal{L}_{k1,k2}$. The discretized operator L induced by $\mathcal{L}_{k1,k2}$ can be given by

$$\mathbf{L} = -i\begin{bmatrix} 0 & I_{N\times N} \\ -16\dfrac{EI}{\rho}D^4 & 0 \end{bmatrix}, \tag{64}$$

where $I_{N\times N}$ is the N×N identity matrix and D is the derivative matrix (58).

5.2 Incorporating the boundary conditions

Discretization of the boundary conditions in the domain description (60) yields

$$\begin{gathered}\Phi_N = 0, \quad [D\Phi]_N = 0, \quad \Theta_N = 0, \quad [D\Theta]_N = 0, \\ 4EI\left[D^2\Phi\right]_1 + k_1\Theta_1 = 0, \quad 4EI\left[D^3\Phi\right]_1 - k_2[D\Theta]_1 = 0.\end{gathered} \tag{65}$$

Let $r_N, z_N, l_N \in \mathbb{R}^N$ be auxiliary row-vectors

$$r_N = [0 \quad 0 \quad \cdots \quad 0 \quad 1], \quad z_N = [0 \quad 0 \quad \cdots \quad 0 \quad 0], \quad l_N = [1 \quad 0 \quad \cdots \quad 0 \quad 0] \tag{66}$$

and D^n_j designate the jth row of the nth derivative matrix D^n. Using Eqs. (66) we represent Eqs. (65) as the following matrix equation:

$$\mathbb{K}\Psi \equiv \begin{bmatrix} r_N & z_N \\ D^1_N & z_N \\ z_N & r_N \\ z_N & D^1_N \\ 4EID^2_1 & k_1 l_N \\ 4EID^3_1 & -k_2 D^1_1 \end{bmatrix} \begin{bmatrix} \Phi \\ \Theta \end{bmatrix} = 0. \tag{67}$$

$\mathbb{K}$ is called the boundary operator. Let $\mathcal{K}_N$ be the kernel of $\mathbb{K}$, i.e., $\mathcal{K}_N=\{v\in \mathbb{R}^{2N}:\mathbb{K} v=0\}$. We have to identify all eigenvalues of the operator $\mathbf{L}$, when its domain is restricted to $\mathcal{K}^N$. It is clear that $\mathcal{K}_N$ is isomorphic to $\mathbb{R}^k$ with $k\equiv\dim\mathcal{K}_N=\dim \mathcal{H}_N-\text{rank}\, \mathbb{K}=2N-6$. Let B be the matrix consisting of column vectors that form an orthonormal basis in $\mathcal{K}_N$. It is clear that B^TB is the identity matrix on $\mathbb{R}^k$ and BB^T is the identity matrix on $\mathbb{K}$. The following result holds: if λ is an eigenvalue of the operator L, and the corresponding eigenvector Ψ satisfies Eq. (67), then the same λ is an eigenvalue of the matrix ($B^T\mathbf{L}B$). However, the inverse statement is not necessarily true. Indeed, we observe that BB^T is the identity in $\mathcal{K}_N$, which is not equivalent to the identity in $\mathcal{H}_N$. Assume now that λ is an eigenvalue of $B^T\mathbf{L}B$ with corresponding eigenvector $v\in\mathbb{R}^k$. If $\Psi=Bv$, we have

$$BB^T\mathbf{L}\Psi = BB^T\mathbf{L}Bv = \lambda Bv = \lambda\Psi, \tag{68}$$

but $BB^T\mathbf{L}\neq\mathbf{L}$, which indicates that fake eigenvalues may exist.

5.3 Filtering of spurious eigenvalues

In order to decide which eigenvalues of B^TLB should be discarded, we impose the following condition. Let Λ be the spectrum of B^TLB and V be the set of its eigenfunctions. We construct the set of "trusted" eigenvalues [14, 15], for some $\varepsilon_f>0$ filtering precision, as

$$\Lambda_\varepsilon = \left\{\lambda\in\Lambda : \|LBv_\lambda - \lambda Bv_\lambda\|_C < \varepsilon_f, \text{ for corresponding eigenvector } v_\lambda\in V\right\}, \tag{69}$$

where $\|\cdot\|_C$ is a discrete approximation to the integral norm defined in Eq. (61). (The subscript C is short for Chebyshev). Using the CGL quadrature, we obtain the following formula for the norm of a vector Ψ defined as in Eq. (63):

$$\|\boldsymbol{\Psi}\|_C = \frac{\pi/4}{N-1}\sum_{k=1}^{N}\sqrt{1-\xi_k^2}\left[16EI(\xi_k)\left|\left[D^2\boldsymbol{\Phi}\right]_k\right|^2 + \rho(\xi_k)\left|\Theta_k\right|^2\right]$$

6. CONCISE SUMMARY

This study examines the spectrum characteristics of the Euler-Bernoulli beam model, specifically focusing on boundary conditions that involve feedback. The model's dynamics generator is a non-self-adjoint matrix differential operator that operates in a Hilbert space consisting of two-component Cauchy data. The operator has been estimated using Chebyshev polynomial approximation, resulting in a discrete operator. We have demonstrated that the eigenvalues of the primary operator may be estimated with great precision using the eigenvalues of its discrete equivalent. Practitioners who require the elastic modes might utilize the leading asymptotic terms in formulae (20) and (21). Additional findings pertain to the presence and equations of the deadbeat modes. Research has demonstrated that when one control parameter, k1, approaches a value slightly greater than 1, and the other parameter, k2, is equal to 0, the number of deadbeat modes increases indefinitely. The formula for the asymptotic growth rate of the number of deadbeat modes has been derived.

It has also been established that there exists a sequence $\left\{k_1^{(n)}\right\}_{n=1}^{\infty}$ of the values of parameter k_1, such that the corresponding deadbeat mode has a multiplicity 2, which yields the existence of the associate mode shapes for the operator $\mathcal{L}_{k1,k2}$. The formulas for the double deadbeat modes and asymptotics for the sequence $\left\{k_1^{(n)}\right\}$ as $n \to \infty$ have been derived.

7. AT A GLANCE

The study has provided both analytic and numerical findings of the Euler-Bernoulli beam model, specifically focusing on a set of boundary conditions characterized by two parameters. The co-diagonal matrix, which is determined by two control parameters (k_1 and k_2), establishes a relationship between a two-dimensional input vector (representing the shear and moment at the right end) and the observation vector (consisting of the time derivatives of displacement and the slope at the right end). The paper contains the following results. Initially, precise numerical estimates for the eigenvalues of the discretized differential operator (which serves as the dynamics generator of the model) have been acquired with a high level of accuracy. Furthermore, the equation for determining the number of deadbeat modes has been derived specifically for the scenario where one control parameter, k1, is positive and the other value, k_2, is zero. It has been demonstrated that the number of deadbeat modes increases indefinitely as k_1 approaches a value slightly more than 1 and k_2 equals 0. Furthermore, the presence of dual deadbeat modes and the asymptotic formula for these modes have been shown. In addition, numerical solutions have been generated to support and confirm all the analytical findings. These results were obtained by employing Chebyshev polynomial approximations for the continuous issue.

REFERENCES

1. Shubov MA, Kindrat LP. Spectral analysis of the Euler–Bernoulli beam model with fully nonconservative feedback matrix. Mathematicsl Methods in the Applied Sciences. 2018;41(12):4691-4713. DOI: 10.1002/mma.4922
2. Shubov MA, Shubov VI. Stability of a flexible structure with destabilizing boundary conditions. Proceedings of the Royal Society A. 2016;472(2191):20160109. DOI: 10.1098/rspa.2016.0109
3. Shubov MA. Exact controllability of nonselfadjoint Euler-Bernoulli beam model via spectral decomposition method. IMA Journal of Mathematical Control and Information. 2008;25(2):185-203. DOI: 10.1093/imamci/dnm018
4. Guiver CH, Opmeer MR. Non-dissipative boundary feedback for Rayleigh and Timoshenko beams. Systems and Control Letters. 2010;59(9):578-586. DOI: 10.1016/j.sysconle.2010.07.002
5. Shubov MA. Exact controllability of coupled Euler-Bernoulli and Timoshenko beam model. IMA Journal of Mathematical Control and Information. 2006;23(3):279-300. DOI: 10.1093/imamci/dnm059
6. Curtain RF, Zwart HJ. An Introduction to Infinite Dimensional Linear Systems Theory. New York, NY: Springer; 1995
7. Dowell EH, editor. A Modern Course on Aeroelasticity. 4th Revised ed. Boston, MA: Klumer Academic Publishers; 2004
8. Balakrishnan AV, Shubov MA. Asymptotic behaviour of aeroelastic modes for aircraft wing model in subsonic air flow. Proceedings of the Royal Society A. 2004;460(2044):1057-1091. DOI: 10.1098/rspa.2003.1217
9. Shubov MA, Rojas-Araneza M. Four-branch vibrational spectrum of double-walled carbon nanotube model. Proceedings of the Royal Society A. 2011;467(2125):99-126. DOI: 10.1098/rspa.2010.0100
10. Benaroya H. Mechanical Vibration: Analysis, Uncertainties, and Control. Upper Saddle River, NJ: Prentice Hall; 1998
11. Han SM, Benaroya H, Wei T. Dynamics of transversely vibrating beams using four engineering theories. Journal of Sound and Vibration. 1999;225(5):935-988. DOI: 10.1006/jsvi.1999.2257

12. Shubov MA. Generation of Gevrey class semigroup by non-selfadjoint Euler-Bernoulli beam model. Mathematicsl Methods in the Applied Sciences. 2006;29(18):2181-2199. DOI: 10.1002/mma.768

13. Gohberg IT, Krein MG. Introduction to the Theory of Nonselfadjoint Operators in Hilbert Space. Translations of Mathematical Monographs No. 18. Providence, RI: AMS; 1996

14. Mason JC, Handscomb DC. Chebyshev Polynomials. Boca Raton, FL: Chapman & Hall/CRC; 2003

15. Trefethen LN, Bau D III. Numerical Linear Algebra. Philadelphia, PA: SIAM; 1997

16. Shubov MA, Wineberg S, Holt R. Numerical investigation of aeroelastic mode distribution for aircraft wing model in subsonic air flow. Mathematical Problems in Engineering. 2010;2010:879519. DOI: 10.1155/2010/879519

17. Marcus AS. Introduction to the Spectral Theory of Polynomial Operator Pencils. Translations of Mathematical Monographs No. 71. Providence, RI: AMS; 1988

18. Fedoryuk MV. Asymptotic Analysis. Berlin, Heidelberg: Springer-Verlag; 1993

19. Naimark MA. Linear Differential Operators. New York, NY: Ungar Pub Co; 1967

10 STUDY OF INEQUALITIES AND EQUATIONS INVOLVING INTEGRALS AND DERIVATIVES USING FRACTIONAL CALCULUS

1. INTRODUCTION: AN OVERVIEW

In recent decades, there has been significant advancement in fractional calculus, mostly driven by its crucial applications across several study domains [1, 2, 3, 4]. Conversely, integral inequalities play a crucial role in probability theory and applied sciences. To obtain additional information, we direct the reader to references [5, 6, 7, 8, 9, 10, 11, 12] and the sources cited within those references. Furthermore, the examination of integral inequalities utilizing fractional integration theory holds significant significance. For specific applications, please refer to references [1, 13, 14, 15, 16, 17].

Boundary value problems of fractional differential equations are a significant focus in the field of fractional calculus. These problems arise in various scientific and engineering applications, such as mechanics, chemistry, electricity, biology, finance, and control theory. To obtain additional information, we direct the reader to references [3, 18, 19, 20].

This chapter employs Riemann-Liouville integrals to demonstrate certain findings pertaining to Minkowski and Hermite-Hadamard inequalities [21]. We illustrate our results by establishing several classes of fractional integral inequalities employing a family of positive functions. These inequalities can be seen as generalizations of some previous fractional and classical integral results that were recently reported [22]. Subsequently, we utilize the generated applications to establish new lower bounds by estimating the fractional expectations and variances for the beta random variable. The traditional covariance identities, which are valid for $\alpha=1$, are extended to any $\alpha\geq1$ and $\beta\geq1$; refer to [23].

In the field of differential equations, my collaborator and I have made a significant contribution. We have developed a class of fractional differential equations that extend the existing chaotic electrical circuit model. We demonstrate recent findings about the presence and singularity of solutions for a specific category of Langevin-type equations. Furthermore, this study examines additional criteria for the data of the problem that guarantee the presence of at least one solution. The class under consideration exhibits a certain association with the well-regarded paper mentioned in reference [20].

The chapter is organized in the following manner: In Section 2, we review several fundamental concepts of fractional calculus that will be utilized in the chapter. Section 3 focuses on the primary findings regarding integral inequalities and also includes other estimations related to continuous random variables. In Section 4, we focus on a specific form of differential equations known as Langevin equations. We investigate the existence and uniqueness of solutions for this class of equations using the Banach contraction principle. Additionally, we utilize the Schaefer fixed point theorem to establish an existence result. The Conclusion comes after everything else has been discussed.

2. PRELIMINARIES ON FRACTIONAL CALCULUS

In this section, we present some definitions and lemmas that will be used in this chapter. For more details, we refer the reader to [2, 13, 15, 24].

Definition 1.1. The Riemann-Liouville fractional integral operator of order $\alpha \geq 0$, for a continuous function f on $[a,b]$ is defined as

$$J_a^{\alpha}[f(t)] = \frac{1}{\Gamma(\alpha)}\int_a^t (t-\tau)^{\alpha-1} f(\tau)d\tau, \alpha > 0, \ a < t \leq b, \tag{1}$$

$$J_a^0[f(t)] = f(t),$$

where $\Gamma(\alpha) := \int_0^{\infty} e^{-u}u^{\alpha-1}du.$

Note that for $\alpha>0$, $\beta>0$, we have

$$J^{\alpha}J^{\beta}f(t) = J^{\alpha+\beta}f(t), \tag{2}$$

and

$$J_a^{\alpha}J_a^{\beta}[f(t)] = J_a^{\beta}J_a^{\alpha}[f(t)]. \tag{3}$$

In the rest of this chapter, for short, we note a probability density function by *p.d.f* . So, let us consider a positive continuous function ω defined on $[a,b]$. We recall the ω- concepts:

Definition 1.2. The fractional ω- weighted expectation of order $\alpha>0$, for a random variable X with a positive *p.d.f.. f* defined on $[a,b]$, is given by

$$E_{\alpha,\omega}(X) := J_a^\alpha[t\omega f](b) = \frac{1}{\Gamma(\alpha)}\int_a^b (b-\tau)^{\alpha-1}\tau\omega(\tau)f(\tau)d\tau, \alpha>0, a<t\le b, \quad (4)$$

Definition 1.3. The fractional ω- weighted variance of order α>0 for a random variable X having a *p.d.f. f* on [a, b] is given by

$$\sigma^2_{\alpha,\omega}(X) = V_{\alpha,\omega}(X) := \frac{1}{\Gamma(\alpha)}\int_a^b (b-\tau)^{\alpha-1}(\tau - E(X))^2\omega(\tau)f(\tau)d\tau, \alpha>0. \quad (5)$$

Definition 1.4. The fractional ω- weighted moment of orders *r*>0, α>0, for a continuous random variable X having a *p.d.f. f* defined on [a, b] is defined by the quantity:

$$E_{\alpha,\omega}(X^r) := \frac{1}{\Gamma(\alpha)}\int_a^b (b-\tau)^{\alpha-1}\tau^r\omega(\tau)f(\tau)d\tau, \alpha>0. \quad (6)$$

We introduce the covariance of fractional order as follows.

Definition 1.5. Let f_1 and f_2 be two continuous on $[a,b]$. We define the fractional ω- weighted covariance of order α>0 for $(f_1(X), f_2(X))$ by

$$Cov_{\alpha,\omega}(f_1(X), f_2(X)) := \frac{1}{\Gamma(\alpha)}\int_a^b (b-\tau)^{\alpha-1}(f_1(\tau)-f_1(\mu))(f_2(\tau)-f_2(\mu))\omega(\tau)f(\tau)d\tau, \alpha>0, \quad (7)$$

where μ is the classical expectation of X.

It is to note that when ω(*x*)=1, *x*∈[*a*,*b*], then we put

$$Var_{\alpha,\omega}(X) := Var_\alpha(X), Cov_{\alpha,\omega}(X) := Cov_\alpha(X), E_{\alpha,\omega}(X) := E_\alpha(X)$$

Definition 1.6. For a function $K\in C^n([a,b],\mathbb{R})$ and $n-1<\alpha\le n$, the Caputo fractional derivative of order α is defined by

$$D^\alpha K(t) = J^{n-\alpha}\frac{d^n}{dt^n}(K(t))$$

$$= \frac{1}{\Gamma(n-\alpha)}\int_a^t (t-s)^{n-\alpha-1}K^{(n)}(s)ds.$$

We recall also the following properties.

Lemma 1.7. Let $n\in\mathbb{N}^*$, and $n-1<\alpha<n$. The general solution of $D^\alpha y(t)=0$, t∈[*a*,*b*] is given by

$$y(t) = \sum_{i=0}^{n-1} c_i(t-a)^i, \quad (8)$$

where $c_i \in \mathbb{R}$, $i=0,1,2,..,n-1$.

Lemma 1.8. Let $n \in \mathbb{N}^*$ and $n-1<\alpha<n$. Then

$$J^\alpha D^\alpha y(t) = y(t) + \sum_{i=0}^{n-1} c_i (t-a)^i, t \in [a,b], \tag{9}$$

for some $c_i \in \mathbb{R}$, $i=0,1,2,..,n-1$.

3. SOME INTEGRAL INEQUALITIES

3.1 On Minkowski and Hermite-Hadamard fractional inequalities

In this subsection, we present some fractional integral results related to Minkowski and Hermite-Hadamard integral inequalities. For more details, we refer the reader to [21].

Theorem 1.9. Let $\alpha>0$, $p\geq1$ and let f,g be two positive functions on $[0,\infty[$, such that for all $t>0, J^\alpha f^p(t)<\infty, J^\alpha g^p(t)<\infty$. If $0<m\leq \frac{f(\tau)}{g(\tau)} \leq M, \tau \in [0,t]$, then we have

$$[J^\alpha f^p(t)]^{\frac{1}{p}} + [J^\alpha g^p(t)]^{\frac{1}{p}} \leq \frac{1+M(m+2)}{(m+1)(M+1)} [J^\alpha (f+g)^p(t)]^{\frac{1}{p}}. \tag{10}$$

Proof: We use the hypothesis $\frac{f(\tau)}{g(\tau)} < M, \tau \in [0,t], t>0$. We can write

$$\begin{aligned} &\frac{(M+1)^p}{\Gamma(\alpha)} \int_0^t (t-\tau)^{\alpha-1} f^p(\tau) d\tau \\ &\leq \frac{M^p}{\Gamma(\alpha)} \int_0^t (t-\tau)^{\alpha-1} (f+g)^p(\tau) d\tau. \end{aligned} \tag{11}$$

Hence, we have

$$J^\alpha f^p(t) \leq \frac{M^p}{(M+1)^p} J^\alpha (f+g)^p(t). \tag{12}$$

Thus, it yields that

$$[J^\alpha f^p(t)]^{\frac{1}{p}} \leq \frac{M}{M+1} [J^\alpha (f+g)^p(t)]^{\frac{1}{p}}. \tag{13}$$

In the same manner, we have

$$\left(1+\frac{1}{m}\right) g(\tau) \leq \frac{1}{m} (f(\tau)+g(\tau)). \tag{14}$$

And then,

$$[J^\alpha g^p(t)]^{\frac{1}{p}} \leq \frac{1}{m+1} [J^\alpha (f+g)^p(t)]^{\frac{1}{p}}. \tag{15}$$

Combining (13) and (15), we achieve the proof. Remark 1.10. Applying the above theorem for α=1, we obtain Theorem 1.2 of [25] on [0,*t*].

With the same arguments as before, we present the following theorem.

Theorem 1.11. Let α>0,*p*≥1 and let *f*,*g* be two positive functions on [0,∞[,such that for all $t>0, J^\alpha f^p(t)<\infty, J^\alpha g^p(t)<\infty$If $0<m\leq\frac{f(\tau)}{g(\tau)}\leq M, \tau\in[0,t]$, then we have

$$[J^\alpha f^p(t)]^{\frac{1}{p}}+[J^\alpha g^p(t)]^{\frac{1}{p}}\leq\frac{1+M(m+2)}{(m+1)(M+1)}[J^\alpha(f+g)^p(t)]^{\frac{1}{p}}. \qquad (16)$$

Remark 1.12. Taking α=1in this second theorem, we obtain Theorem 2.2 in [26] on [0,*t*].

Using the notions of concave and **L^p**- functions, we present to the reader the following result.

Theorem 1.13. Suppose that α>0, *p*>1, *q*>1 and let *f*,*g* be two positive functions on [0,∞[. If f^p, g^q are two concave functions on [0,∞[, then we have

$$\begin{aligned}&2^{-p-q}(f(0)+f(t))^p(g(0)+g(t))^q(J^\alpha(t^{\alpha-1}))^2\\&\leq J^\alpha(t^{\alpha-1}f^p(t))J^\alpha(t^{\alpha-1}g^q(t)).\end{aligned} \qquad (17)$$

The proof of this theorem is based on the following auxiliary result.

Lemma 1.14. Let *h* be a concave function on [*a*,*b*]. Then for any *x*∈[*a*,*b*], we have

$$h(a)+h(b)\leq h(b+a-x)+h(x)\leq 2h\left(\frac{a+b}{2}\right). \qquad (18)$$

3.2 A family of fractional integral inequalities

We present to the reader some integral results for a family of functions [22]. These results generalize some integral inequalities of [27]. We have

Theorem 1.15. Suppose that $(f_i)_{i=1,\dots,n}$ are *n* positive, continuous, and decreasing functions on [*a*,*b*]. Then, the following inequality

$$\frac{J^\alpha\left[\prod_{i\neq p}^n f_i^{\gamma_i} f_p^{\beta}(t)\right]}{J^\alpha\left[\prod_{i=1}^n f_i^{\gamma_i}(t)\right]}\geq\frac{J^\alpha\left[(t-a)^\delta\prod_{i\neq p}^n f_i^{\gamma_i} f_p^{\beta}(t)\right]}{J^\alpha\left[(t-a)^\delta\prod_{i=1}^n f_i^{\gamma_i}(t)\right]} \qquad (19)$$

holds for any *a*<*t*≤*b*, α>0, δ>0, $\beta\geq\gamma_p>0$, where *p* is a fixed integer in {1,2,...,*n*}.

Proof: It is clear that

$$\left((\rho-a)^\delta-(\tau-a)^\delta\right)\left(f_p^{\beta-\gamma_p}(\tau)-f_p^{\beta-\gamma_p}(\rho)\right)\geq 0, \qquad (20)$$

for any fixed *p*∈{1,...*n*} and for any $\beta\geq\gamma_p>0, \delta>0, \tau,\rho\in[a,t]; a<t\leq b$.

Taking

$$K_p(\tau,\rho) := \frac{(t-\tau)^{\alpha-1}}{\Gamma(\alpha)}\prod_{i=1}^{n} f_i^{\gamma_i}(\tau)\left((\rho-a)^{\delta}-(\tau-a)^{\delta}\right)\left(f_p^{\beta-\gamma_p}(\tau)-f_p^{\beta-\gamma_p}(\rho)\right), \quad (21)$$

we observe that

$$K_p(\tau,\rho) \geq 0. \quad (22)$$

Also, we have

$$0 \leq \int_a^t K_p(\tau,\rho)d\tau = (\rho-a)^{\delta} J^{\alpha}\left[\prod_{i\neq p}^{n} f_i^{\gamma_i} f_p^{\beta}(t)\right] + f_p^{\beta-\gamma_p}(\rho) J^{\alpha}\left[(t-a)^{\delta}\prod_{i=1}^{n} f_i^{\gamma_i}(t)\right]$$
$$-J^{\alpha}\left[(t-a)^{\delta}\prod_{i\neq p}^{n} f_i^{\gamma_i} f_p^{\beta}(t)\right] - (\rho-a)^{\delta} f^{\beta-\gamma_p}(\rho) J^{\alpha}\left[\prod_{i=1}^{n} f_i^{\gamma_i}(t)\right]. \quad (23)$$

Hence, we get

$$J^{\alpha}\left[(t-a)^{\delta}\prod_{i=1}^{n} f_i^{\gamma_i}(t)\right] J^{\alpha}\left[\prod_{i\neq p}^{n} f_i^{\gamma_i} f_p^{\beta}(t)\right] \geq J^{\alpha}\left[\prod_{i=1}^{n} f_i^{\gamma_i}(t)\right] J^{\alpha}\left[(t-a)^{\delta}\prod_{i\neq p}^{n} f_i^{\gamma_i} f_p^{\beta}(t)\right]. \quad (24)$$

The proof is thus achieved.

Remark 1.16. Applying Theorem 1.15 for α=1, *t*=*b*, *n*=1, we obtain Theorem 3 in [27].

Using other sufficient conditions, we prove the following generalization.

Theorem 1.17. Suppose that $(f_i)_{i=1,\dots,n}$ are positive, continuous, and decreasing functions on [*a*,*b*]. Then for any fixed *p* in {1,2,...,*n*} and for any $a < t \leq b, \alpha > 0, \omega > 0, \delta > 0, \beta \geq \gamma_p > 0$, we have

$$\frac{J^{\alpha}\left[\prod_{i\neq p}^{n} f_i^{\gamma_i} f_p^{\beta}(t)\right] J^{\omega}\left[(t-a)^{\delta}\prod_{i=1}^{n} f_i^{\gamma_i}(t)\right] + J^{\omega}\left[\prod_{i\neq p}^{n} f_i^{\gamma_i} f_p^{\beta}(t)\right] J^{\alpha}\left[(t-a)^{\delta}\prod_{i=1}^{n} f_i^{\gamma_i}(t)\right]}{J^{\alpha}\left[(t-a)^{\delta}\prod_{i\neq p}^{n} f_i^{\gamma_i} f_p^{\beta}(t)\right] J^{\omega}\left[\prod_{i=1}^{n} f_i^{\gamma_i}(t)\right] + J^{\omega}\left[(t-a)^{\delta}\prod_{i\neq p}^{n} f_i^{\gamma_i} f_p^{\beta}(t)\right] J^{\alpha}\left[\prod_{i=1}^{n} f_i^{\gamma_i}(t)\right]} \geq 1. \quad (25)$$

Proof: Multiplying both sides of (23) by $\frac{(t-\rho)^{\omega-1}}{\Gamma(\omega)}\prod_{i=1}^{n} f_i^{\gamma_i}(\rho), \omega > 0,$ then integrating the resulting inequality with respect to ρ over (*a*,*t*),*a*<*t*≤*b* and using Fubini's theorem, we obtain the desired inequality.

Remark 1.18.

(i) Applying Theorem 1.17 for α=ω, we obtain Theorem 1.15.

(ii) Applying Theorem 1.17 for α=ω=1, *t*=*b*, *n*=1, we obtain Theorem 3 of [27].

Introducing a positive increasing function g to the family $(f_i)i=1,\dots n$ we establish the following theorem.

Theorem 1.19. Let $(f_i)_{i=1,\dots n}$ and g be positive continuous functions on $[a,b]$, such that g is increasing and $(f_i)_{i=1,\dots n}$ are decreasing on $[a,b]$. Then, the following inequality

$$\frac{J^{\alpha}\left[\prod_{i\neq p}^{n} f_i^{\gamma_i} f_p^{\beta}(t)\right] J^{\alpha}\left[g^{\delta}(t)\prod_{i=1}^{n} f_i^{\gamma_i}(t)\right]}{J^{\alpha}\left[g^{\delta}(t)\prod_{i\neq p}^{n} f_i^{\gamma_i} f_p^{\beta}(t)\right] J^{\alpha}\left[\prod_{i=1}^{n} f_i^{\gamma_i}(t)\right]} \geq 1 \qquad (26)$$

holds for any $a<t\leq b$, $\alpha>0$, $\delta>0$, $\beta\geq\gamma p>0$, where p is a fixed integer in $\{1,2,\dots,n\}$.

Remark 1.20. Applying Theorem 1.19 for $\alpha=1$, $t=b, n=1$, we obtain Theorem 4 of [27].

3.3 Some estimations on random variables

3.3.1 Bounds for fractional moments of beta distribution

In what follows, we present some fractional results on the beta distribution [23]. So let us prove the following α- version.

Theorem 1.21. Let X,Y,U and V be four random variables, such that $X\sim B(p,q)$, $Y\sim B(m,n)$, $U\sim B(p,n)$, and $V\sim B(m,q)$. If $(p-m)(q-n)\leq 0$, then

$$\frac{E_{\alpha}(X^r)E_{\alpha}(Y^r)}{E_{\alpha}(U^r)E_{\alpha}(V^r)} \geq \frac{B(p,n)B(m,q)}{B(p,q)B(m,n)}, \alpha\geq 1.$$

For the proof of this result, we can apply a weighted version of the fractional Chebyshev inequality as is mentioned in [1].

Remark 1.22. The above theorem generalizes Theorem 3.1 of [7].

We propose also the following (α,β)-version that generalizes the above result. We have

Theorem 1.23. Let X,Y,U, and V be four random variables, such that $X\sim B(p,q)$, $Y\sim B(m,n)$, $U\sim B(p,n)$, and $V\sim B(m,q)$. If $(p-m)(q-n)\leq 0$, then

$$\frac{E_{\alpha}(X^r)E_{\beta}(Y^r)+E_{\beta}(X^r)E_{\alpha}(Y^r)}{E_{\alpha}(U^r)E_{\beta}(V^r)+E_{\beta}(U^r)E_{\alpha}(V^r)} \geq \frac{B(p,n)B(m,q)}{B(p,q)B(m,n)}, \alpha,\beta\geq 1.$$

Remark 1.24. If $\alpha=\beta=1$, then the above theorem reduces to Theorem 3.1 of [7].

3.3.2 Identities and lower bounds

In the following theorem, the fractional covariance of X and $g(X)$ is expressed with the derivative of $g(X)$. It can be considered as a generalization of a covariance identity established by the authors of [28]. So, we prove the result:

Theorem 1.25. Let X be a random variable having a *p.d.f* defined on $[a,b]$; $\mu=E(X)$.

Then, we have

$$Cov_{\alpha}(X,g(X)) = \frac{1}{\Gamma(\alpha)}\int_a^b g'(x)dx\int_a^x (b-t)^{\alpha-1}(\mu-t)f(t)dt, \alpha\geq 1. \quad (27)$$

We can prove this result by the application of the covariance definition in the case where $\omega(x)$=1.

The following theorem establishes a lower bound for $Var_{\alpha}(g(X))$ of any function $g\in C^1([a,b])$. We have

Theorem 1.26. Let X be a random variable having a *p.d.f* defined on $[a,b]$, such that $\mu=E(X)$. Then, we have

$$Var_{\alpha}(g(X)) \geq \frac{1}{Var_{X,\alpha}}\left(\frac{1}{\Gamma(\alpha)}\int_a^b g'(x)dx\int_a^x (b-t)^{\alpha-1}(\mu-t)f(t)dt\right)^2, \quad (28)$$

for any $g\in C^1([a,b])$.

To prove this result, we use fractional Cauchy-Schwarz inequality established in [29].

Remark 1.27. Let us consider $\Omega\in C([a,b])$ that satisfies $\int_a^x (b-t)^{\alpha-1}(\mu-t)f(t)dt = (b-x)^{\alpha-1}\sigma^2\Omega(x)f(x)$. Then, we present the following result.

Theorem 1.28. Let X be a random variable having a *p.d.f.* defined on $[a,b]$, such that $\mu = E(X), \sigma^2 = Var(X)$ and $\Omega\in C([a,b])$; $\int_a^x (b-t)^{\alpha-1}(\mu-t)f(t)dt = (b-x)^{\alpha-1}\sigma^2\Omega(x)f(x)$.

Then, we have

$$Var_{\alpha}(g(X)) \geq \frac{\sigma^4(X)}{Var_{\alpha}(X)}E_{\alpha}^2(g'(X)\Omega(X)). \quad (29)$$

Proof: We have

$$Cov_{\alpha}^2(X,g(X)) = \left[\frac{1}{\Gamma(\alpha)}\int_a^b g'(x)dx(b-x)^{\alpha-1}\sigma^2\Omega(x)f(x)dx\right]^2. \quad (30)$$

On the other hand, we can see that

$$\left[\frac{1}{\Gamma(\alpha)}\int_a^b g'(x)dx(b-x)^{\alpha-1}\sigma^2\Omega(x)f(x)dx\right]^2 = \sigma^4 E_{\alpha}^2(g'(X)\Omega(X)) \quad (31)$$

Thanks to the fractional version of Cauchy Schwarz inequality [29], and using the fact that

$$Cov_{\alpha}^2(X,g(X)) \leq Var_{\alpha}(X)Var_{\alpha}(g(X)), \quad (32)$$

we obtain

$$\sigma^4 E_\alpha^2(g'(X)\Omega(X)) \leq Var_\alpha(X)Var_\alpha(g(X)). \tag{33}$$

This ends the proof.

Remark 1.29. Thanks to (30) and (31), we obtain the following fractional covariance identity

$$\sigma^2 E_\alpha(g'(X)\Omega(X)) = Cov_\alpha(X, g(X)).$$

It generalizes the good standard identity obtained in [28] that corresponds to $\alpha=1$ and it is given by

$$\sigma^2 E(g'(X)\Omega(X)) = Cov(X, g(X)).$$

We end this section by proving the following fractional integral identity between covariance and expectation in the fractional case.

Theorem 1.30. Let X be a continuous random variable with a *p.d.f.* having a support an interval $[a,b]$, $E(X)=\mu$. Then, for any $\alpha\geq 1$, the following general covariance identity holds

$$Cov_\alpha(h(X), g(X)) = E_\alpha(g'(X)Z(X)), \tag{34}$$

where $g\in C^1([a,b])$, with $E|Z(X)g'(X)|<\infty$, $h(x)$ is a given function and $Z(x)f(x)\frac{(b-x)^{\alpha-1}}{\Gamma(\alpha)} = \int_a^x (E(h(X)) - h(t))\frac{(b-t)^{\alpha-1}}{\Gamma(\alpha)}f(t)dt.$

Proof: We have

$$Cov_\alpha(h(X), g(X)) = \frac{1}{\Gamma(\alpha)}\int_a^b (b-x)^{\alpha-1}(h(x) - h(\mu))(g(x) - g(\mu))f(x)dx \tag{35}$$

and

$$E_\alpha(g'(X)Z(X)) = \frac{1}{\Gamma(\alpha)}\int_a^b (b-x)^{\alpha-1}g'(x)Z(x)f(x)dx. \tag{36}$$

The definition of $Z(X)$ implies that

$$E_\alpha(g'(X)Z(X)) = \frac{1}{\Gamma(\alpha)}\int_a^\mu (g(\mu) - g(t))(b-t)^{\alpha-1}(h(\mu) - h(t))f(t)dt \tag{37}$$

$$+\frac{1}{\Gamma(\alpha)}\int_\mu^b (g(t) - g(\mu))(b-t)^{\alpha-1}(h(t) - h(\mu))f(t)dt.$$

Hence, we obtain

$$E_\alpha(Z(X)g'(X)) = Cov_\alpha(g(X), h(X)). \tag{38}$$

Remark 1.31. Taking $\alpha=1$, in the above theorem, we obtain Theorem 2.2 of [10].

4. A CLASS OF DIFFERENTIAL EQUATIONS OF FRACTIONAL ORDER

Inspired by the work in [4, 20], in what follows we will be concerned with a more general class of Langevin equations of fractional order. The considered class will contain a nonlinearity that depends on a fractional derivative of order δ. So, let us consider the following problem:

$$\begin{aligned} &{}^cD^{\alpha}(D^2+\lambda^2)u(t) = f(t, u(t), {}^cD^{\delta}u(t)), \\ &t \in [0,1], \quad \lambda \in \mathbb{R}_+^* \\ &0 < \alpha \le 1, \quad 0 \le \delta < \alpha, \end{aligned} \tag{39}$$

associated with the conditions

$$u(0) = 0, \quad u''(0) = 0, \quad u(1) = \beta u(\eta), \eta \in (0,1), \tag{40}$$

where ${}^cD^{\alpha}$ denotes the Caputo fractional derivative of fractional order α, D^2 is the two-order classical derivative, f:[0,1]× ℝ × ℝ → ℝ is a given function, and β∈ℝ, such that $\beta sin(\lambda\eta) \neq sin(\lambda)$.

4.1 Integral representation

We recall the following result [20]:

Lemma 1.32. Let θ be a continuous function on [0,1] . The unique solution of the problem

$$\begin{aligned} &{}^cD^{\alpha}(D^2+\lambda^2)u(t) = \theta(t), \\ &t \in [0,1], \quad \lambda \in \mathbb{R}_+^* \\ &n-1 < \alpha \le n, \quad n \in \mathbb{N}^*, \end{aligned} \tag{41}$$

is given by

$$u(t) = \frac{1}{\lambda}\int_0^t sin\lambda(t-s)\left(\int_0^s \frac{(s-\tau)^{\alpha-1}}{\Gamma(\alpha)}\theta(\tau)d\tau + \sum_{i=1}^{n-1} c_i s^i\right)ds + c_n\, cos(\lambda t) + c_{n+1} sin(\lambda t), \tag{42}$$

where $c_i \in \mathbb{R}$, $i=1...n+1$.

Thanks to the above lemma, we can state that

The class of Langevin equations (39) and (40) has the following integral representation:

$$u(t) = \frac{1}{\lambda}\int_0^t sin\lambda(t-s)\left(\int_0^s \frac{(s-\tau)^{\alpha-1}}{\Gamma(\alpha)} f(\tau, u(\tau), D^{\delta}(\tau))d\tau\right)ds$$
$$+\frac{\sin(\lambda t)}{\Delta}\left[\beta\int_0^{\eta} sin\lambda(\eta-s)\left(\int_0^s \frac{(s-\tau)^{\alpha-1}}{\Gamma(\alpha)} f(\tau, u(\tau), D^{\delta}(\tau))d\tau\right)ds\right.$$
$$\left.-\int_0^1 sin\lambda(1-s)\left(\int_0^s \frac{(s-\tau)^{\alpha-1}}{\Gamma(\alpha)} f(\tau, u(\tau), D^{\delta}(\tau))d\tau\right)ds\right], \tag{43}$$

where

$$\Delta := \lambda(sin\lambda - \beta sin\lambda\eta). \tag{44}$$

4.2 Existence and uniqueness of solutions

Using the above integral representation (43), we can prove the following existence and uniqueness theorem.

Theorem 1.33. Assume that the following hypotheses are valid:

(H1): The function f:[0,1]× $\mathbb{R}\times\mathbb{R}\rightarrow\mathbb{R}$ is continuous, and there exist two constants $\Lambda_1, \Lambda_2 > 0$, such that for all $t \in [0,1]$ and $u_i, v_i \in \mathbb{R}$, $i=1,2$,

$$|f(t, u_1, u_2) - f(t, v_1, v_1)| \leq \Lambda_1|u_1 - v_1| + \Lambda_2|u_2 - v_2|. \tag{45}$$

(H2): Suppose that $\Lambda \leq \frac{1}{(\Phi+\Upsilon)}$,

where

$$\Phi := \frac{\Delta_1 + \lambda + \beta_1\lambda\eta^{\alpha+1}}{\Gamma(\alpha+2)\lambda\Delta_1}, \Psi := \frac{\Delta_1 s^{\alpha}(\alpha+1) + \lambda^2 + \beta_1\lambda^2\eta^{\alpha+1}}{\Gamma(\alpha+2)\lambda\Delta_1}, \Upsilon := \frac{\Psi}{\Gamma(2-\delta)},$$

$$\Lambda := max(\Lambda_1, \Lambda_2), \quad \Delta_1 = |\Delta|, \quad \beta_1 = |\beta|.$$

Then problem (39) and (40) has a unique solution on [0,1].

Proof: We introduce the space

$$E = \{u; u \in C([0,1]), D^{\delta}u \in C([0,1])\},$$

endowed with the norm $\|u\|_E := \|u\|_{\infty} + \|D^{\delta}u\|_{\infty}$.

Then, $(E, \|.\|_E)$ is a Banach space.

Also, we consider the operator $T: E \rightarrow E$ defined by

$$(Tu)(t) := \frac{1}{\lambda}\int_0^t \sin\lambda(t-s) J_0^\alpha f\left(s, u(s), D^\delta u(s)\right) ds$$

$$+\frac{\sin(\lambda t)}{\Delta}\left[\beta\int_0^\eta \sin\lambda(\eta-s) J_0^\alpha f(s, u(s), D^\delta(s)) ds \right. \tag{46}$$

$$\left. -\int_0^1 \sin\lambda(1-s) J_0^\alpha f(s, u(s), D^\delta u(s)) ds\right]$$

We shall prove that the above operator is contractive over the space E.

Let $u_1, u_2 \in E$. Then, for each $t \in [0,1]$, we have

$$|Tu_1(t) - Tu_2(t)| \leq \frac{1}{\lambda}\int_0^t |\sin\lambda(t-s)| \quad J_0^\alpha |f(s, u_1(s), D^\delta u_1(s)) - f(s, u_2(s), D^\delta u_2(s))| ds$$

$$+\frac{|\sin(\lambda t)|}{|\Delta|}\left[|\beta|\int_0^\eta |\sin\lambda(\eta-s)| \quad J_0^\alpha |f(s, u_1(s), D^\delta u_1(s)) - f(s, u_2(s), D^\delta u_2(s))| ds\right.$$

$$\left.+\int_0^1 |\sin\lambda(1-s) \quad |J_0^\alpha f(s, u_1(s), D^\delta u_1(s)) - f(s, u_2(s), D^\delta u_2(s))| ds\right] := \mathcal{A}$$

By (H1), we have

$$\mathcal{A} \leq \frac{\Lambda}{\Gamma(\alpha+2)}\left(\frac{1}{\lambda} + \frac{1}{|\Delta|} + \frac{|\beta|}{|\Delta|}\eta^{\alpha+1}\right)\left(|u_1 - u_2| + |D^\delta u_1 - D^\delta u_2|\right).$$

Hence, it yields that

$$\|Tu_1 - Tu_2\|_\infty \leq \Lambda\Phi\|u_1 - u_2\|_E. \tag{47}$$

With the same arguments as before, we can write.

$$|T'u_1(t) - T'u_2(t)| \leq \frac{1}{\lambda}|sin\lambda(t-s)||J_0^\alpha|f(s,u_1(s),D^\delta u_1(s)) - f(s,u_2(s),D^\delta u_2(s))|ds$$

$$+\frac{|\lambda cos(\lambda t)|}{|\Delta|}\left[|\beta|\int_0^\eta |sin\lambda(\eta-s)|J_0^\alpha|f(s,u_1(s),D^\delta u_1(s)) - f(s,u_2(s),D^\delta u_2(s))|ds\right.$$

$$\left.+\int_0^1 |sin\lambda(1-s)|J_0^\alpha|f(s,u_1(s),D^\delta u_1(s)) - f(s,u_2(s),D^\delta u_2(s))|ds\right] := \mathcal{B}.$$

Again, by (H1), we obtain

$$\mathcal{B} \leq \Lambda\left(\frac{\Delta_1 s^\alpha(\alpha+1) + \lambda^2 + \beta_1\lambda^2\eta^{\alpha+1}}{\Gamma(\alpha+2)\lambda\Delta_1}\right)\left(|u_1-u_2| + |D^\delta u_1 - D^\delta u_2|\right).$$

Consequently, we get

$$\|T'u_1 - T'u_2\|_\infty \leq \Lambda\Psi\|u_1-u_2\|_E.$$

This implies that

$$\|D^\delta Tu_1 - D^\delta Tu_2\|_\infty \leq \Lambda\Upsilon\|u_1-u_2\|_E. \tag{48}$$

Using (47) and (48), we can state that

$$\|Tu_1 - Tu_2\|_E \leq \Lambda(\Phi+\Upsilon)\|u_1-u_2\|_E.$$

Thanks to (H2), we can say that the operator T is contractive.

Hence, by Banach fixed point theorem, the operator has a unique fixed point which corresponds to the unique solution of our Langevin problem.

4.3 Existence of solutions

We prove the following theorem.

Theorem 1.34. Assume that the following conditions are satisfied:

(H3): The function f:$[0,1]\times\mathbb{R}\times\mathbb{R}\rightarrow\mathbb{R}$ is jointly continuous.

(H4): There exists a positive constant $\mathcal{M}$;$|f(t,u,v)|\leq\mathcal{M}$ for any $t\in[0,1]$, $u,v\in\mathbb{R}$.

Then the problem (39), (40) has at least one solution on [0,1].

Proof: We use Schaefer fixed point theorem to prove this result. So we proceed into three steps.

Step 1: We prove that T is continuous and bounded.

Since the function f is continuous by (H3), then the operator is also continuous; this proof is trivial and hence it is omitted.

Let $\Omega\subset E$ be a bounded set. We need to prove that $T(\Omega)$ is a bounded set.

Let $u \in \Omega$. Then, for any $t \in [0,1]$, we have

$$|(Tu)(t)| \leq \left(\frac{1}{\lambda} + \frac{1}{|\Delta|}\right) \int_0^1 J_0^\alpha |f(s, u(s), D^\delta u(s))| ds + \frac{|\beta|}{|\Delta|} \int_0^\eta J_0^\alpha |f(s, u(s), D^\delta (s))| ds := C$$

Using (H4), we get

$$\|Tu\|_\infty \leq \Phi \mathcal{M}. \tag{49}$$

In the same manner, we find that

$$\|D^\delta Tu\|_\infty \leq \Upsilon \mathcal{M}. \tag{50}$$

From (49) and (50), we have

$$\|Tu\|_E \leq (\Phi + \Upsilon)\mathcal{M}.$$

The operator is thus bounded.

Step 2: Equicontinuity.

Let $u \in E$. Then, for each $t_1, t_2 \in [0,1]$, we have

$$\begin{aligned} |Tu(t_2) - Tu(t_1)| \leq |\frac{1}{\lambda}\Bigg[&\int_0^{t_2} \sin\lambda(t_2 - s) J_0^\alpha f(s, u(s), D^\delta u(s)) ds - \int_0^{t_1} \sin\lambda(t_1 - s) J_0^\alpha f(s, u(s), D^\delta u(s)) ds\Bigg] \\ &+ \frac{\sin(\lambda t_2) - \sin(\lambda t_1)}{\Delta}\Bigg[\beta \int_0^\eta \sin\lambda(\eta - s) J_0^\alpha f(s, u(s), D^\delta u(s)) ds \\ &+ \int_0^1 \sin\lambda(1 - s) J_0^\alpha f(s, u(s), D^\delta u(s)) ds\Bigg] | \\ \leq \frac{|\Theta|}{|\Delta|} |\sin(\lambda t_2) - \sin(\lambda t_1)| &+ \frac{1}{\lambda}\int_{t_1}^{t_2} \sin\lambda(t_2 - s) J_0^\alpha |f(s, u(s), D^\delta u(s))| ds \\ &+ \frac{1}{\lambda}\int_0^{t_2} |(\sin\lambda(t_2 - s) - \sin\lambda(t_1 - s)| J_0^\alpha |f(s, u(s), D^\delta u(s))| ds, \end{aligned} \tag{51}$$

where

$$\Theta := \beta \int_0^\eta |\sin\lambda(\eta - s) J_0^\alpha| f(s, u(s), D^\delta u(s))| ds + \int_0^1 |\sin\lambda(1 - s)| J_0^\alpha| f(s, u(s), D^\delta u(s))| ds.$$

Analogously, we can obtain

$$|T'u(t_2) - T'u(t_1)| \leq \lambda \frac{|\Theta|}{|\Delta|} |cos(\lambda t_2) - cos(\lambda t_1)| + \frac{1}{\lambda} |sin\lambda(t_2 - s) - sin\lambda(t_1 - s)| J_0^{\alpha} |f(s, u(s), D^{\delta}u(s))|.$$

Consequently, we can write

$$|D^{\delta}Tu(t_2) - D^{\delta}Tu(t_1)| \leq J^{1-\delta}|T'u(t_2) - T'u(t_1)| \tag{52}$$

As $t_1 \to t_2$, the right-hand sides of (51) and (52) tend to zero.

Therefore,

$$\|Tu(t_2) - Tu(t_1)\|_E \to 0.$$

The operator T is thus equicontinuous.

As a consequence of Step 1 and Step 2 and thanks to Arzela-Ascoli theorem, we conclude that T is completely continuous.

Step 3: We prove that $\Sigma := \{u \in E; u = \lambda Tu, 0 < \lambda < 1\}$ is a bounded set.

Let $u \in \Sigma$. Then, for each $t \in [0,1]$, the following two inequalities are valid:

$$|u(t)| = |\lambda Tu(t)| \leq |Tu(t)| \leq \mathcal{M}\Phi$$

and

$$|D^{\delta}u(t)| = |\lambda D^{\delta}Tu(t)| \leq |D^{\delta}Tu(t)| \leq \mathcal{M}\Upsilon.$$

Therefore,

$$\|u\|_E \leq \mathcal{M}(\Upsilon + \Phi).$$

Thanks to steps 1, 2, and 3 and by Schaefer fixed point theorem, the operator T has at least one fixed point. This ends the proof of the above theorem.

5. CONCISE SUMMARY

This chapter applies fractional calculus to certain categories of integral inequalities. Indeed, the utilization of the Riemann-Liouville integral has led to the establishment of certain Minkowski and Hermite-Hadamard-type inequalities. Additionally, a number of alternative fractional integral outcomes have been derived, which pertain to a set of positive functions. The resulting results extend several traditional integral inequalities found in the existing literature. This chapter also includes applications involving continuous random variables. We have introduced new identities and discussed various estimates. The study has examined the existence and uniqueness of solutions for a nonlocal boundary value problem involving Langevin equations with two fractional parameters. We have employed the Caputo technique in conjunction with the Banach contraction principle to establish the existence and uniqueness of the result. Furthermore, the use of the Schaefer fixed point theorem has yielded an additional proof of existence. Our technique is easily applicable to a wide range of real-world challenges.

6. AT A GLANCE

This chapter utilizes fractional calculus to establish findings on integral inequalities and differential equations. We have derived several findings pertaining to the Hermite-Hadamard inequality. Next, we derive additional fundamental outcomes associated with the Minkowski inequality. We further demonstrate our findings by introducing novel categories of fractional integral inequalities utilizing a set of positive functions. These categories of inequalities can be regarded as extensions of order n for previously known classical/fractional integral outcomes. As a practical use of inequalities, we create new minimum values that estimate the fractional expectancies and variances for the beta random variable. The classical covariance identities are extended to arbitrary values of $\alpha \geq 1$ and $\beta \geq 1$. In the field of differential equations, we offer a contribution that enables the development of a class of fractional chaotic electrical circuits. We demonstrate recent findings about the presence and singularity of solutions for a specific category of Langevin-type problem. Furthermore, the paper examines the presence of at least one solution by establishing certain adequate requirements.

REFERENCES

1. Belarbi S, Dahmani Z. On some new fractional integral inequalities. Journal of Inequalities in Pure and Applied Mathematics. 2009;10(3):1-12
2. Kilbas AA, Srivastava M, Trujillo JJ. Theory and Applications of Fractional Differential Equations. Amsterdam: Elsevier Science B.V; 2006
3. Podlubny I. Fractional Differential Equations: An Introduction to Fractional Derivatives, Fractional Differential Equations, to Methods of Their Solution and Some of Their Applications. San Diego: Academic Press; 1999
4. Salem A, Alzahrani F, Almaghamsi L. Fractional Langevin equation with nonlocal integral boundary condition. Mathematics. 2019;7:110
5. Barnett NS, Cerone P, Dragomir SS, Roumeliotis J. Some inequalities for the expectation and variance of a random variable whose PDF is n-time diferentiable. Journal of Inequalities in Pure and Applied Mathematics. 2000;1(21):1-29
6. Barnett NS, Cerone P, Dragomir SS, Roumeliotis J. Some inequalities for the dispersion of a random variable whose PDF is defined on a finite interval. Journal of Inequalities in Pure and Applied Mathematics. 2001;2(1):1-18
7. Kumar P, Singh SP, Dragomir SS. Some inequalities involving beta and gamma functions. Nonlinear Analysis Forum. 2001;6(1):143-150
8. Kumar P. Moment inequalities of a random variable defined over a finite interval. Journal of Inequalities in Pure and Applied Mathematics. 2002;3(3):1-24

9. Kumar P. Inequalities involving moments of a continuous random variable defined over a finite interval. Computers and Mathematics with Applications. 2004;48:257-273
10. Mohtashami Borzadaran GR, Shanbhag DN. Further results based on Chernoff-type inequalities. Statistics & Probability Letters. 1998;39:109-117
11. Niezgoda M. New bounds for moments of continuous random variables. Computers & Mathematcs with Applications. 2010;60(12):3130-3138
12. Sharma R, Devi S, Kapoor G, Ram S, Barnett NS. A brief note on some bounds connecting lower order moments for random variables defined on a finite interval. International Journal of Applied Science and Research. 2009;1(2):83-85
13. Dahmani Z. Fractional integral inequalities for continuous random variables. Malaya Journal of Matematik. 2014;2(2):172-179
14. Dahmani Z, Tabharit L. On weighted Gruss type inequalities via fractional integrals. Journal of Advanced Research in Pure Mathematics. 2010;2(4):31-38
15. Dahmani Z, Bouziane AE, Houas M, Sarikaya MZ. New w-weighted concepts for continuous random variables with applications. Note di Matematica. 2017;37(1):2340
16. Sarikaya MZ, Yaldiz H. New generalization fractional inequalities of Ostrowski-Gruss type. Lobachevskii Journal of Mathematics. 2013;34(4):326-331
17. Sarikaya MZ, Yaldiz H, Basak N. New fractional inequalities of Ostrowski-Gruss type. Le Matematiche. 2014;69:227-235
18. Dahmani M, Anber A, Dahmani Z. Speech movements on vocal tract: Fractional nonlinear dynamics. Journal of Information and Optimization Sciences. 2019;40(6):1307-1315
19. Mainradi F, Pironi P. The fractional Langevin equation: Brownian motion revisited. Extracta Mathematicae. 1996;11(1):140-154
20. Salem A, Alzahrani F, Almaghamsi L. Langevin equation involving one fractional order with three point boundary conditions. Journal of Nonlinear Sciences and Applications. 2019;12:791-798
21. Dahmani Z. On Minkowski and Hermite-Hadamard integral inequalities via fractional integration. Annals of Functional Analysis. 2010;1(1):51-58
22. Dahmani Z. New classes of integral inequalities of fractional order. Le Matematiche. 2014;LXIX(1):237-247

23. Dahmani Z. New identities and lower bounds for CUD and beta distributions. ROMAI Journal. 2019

24. Gorenflo R, Mainardi F. Fractional Calculus: Integral and Differential Equations of Fractional Order. Wien: Springer Verlag; 1997. pp. 223-276

25. Bougoffa L. On Minkowski and Hardy integral inequality. Journal of Inequalities in Pure and Applied Mathematics. 2003;7(2):60

26. Set E, Ozdemir ME, Dragomir SS. On the Hermite-Hadamard inequality and other integral inequalities involving two functions. Journal of Inequalities and Applications. 2010:148102. DOI: 10.1155/2010/148102

27. Liu W, Ngo QA, Huy VN. Several interesting integral inequalities. Journal of Mathematical Inequalities. 2009;10(2):201-212

28. Caccoulos T, Papathanasiou V. Characterizations of distribution by variance bounds. Statistics & Probability Letters. 1989;7:351-356

29. Dahmani Z. About some integral inequalities using Riemann-Liouville integrals. General Mathematics. 2012;20(4):63-69

11 OBTAINING APPROXIMATE SOLUTIONS FOR CERTAIN BOUNDARY VALUE PROBLEMS USING OPERATIONAL MATRICES OF BERNSTEIN POLYNOMIALS

1. INTRODUCTION: AN OVERVIEW

Fractional calculus, a generalization of classical calculus, is a rapidly expanding field of research. One particularly important aspect is the theory of fractional order differential equations, which has a broad range of applications in real-world issues. Fractional order differential equations offer a highly effective method for describing a wide range of physical and biological processes. The aforementioned equations are crucial for elucidating the hereditary traits of different materials and genetic issues in biological models, in contrast to integer order differential equations in the form of mathematical models. Currently, the majority of its uses are in the field of bio-medical engineering, as well as in other scientific and engineering areas such as mechanics, chemistry, viscoelasticity, control theory, signal and image processing, economics, and optimization theory. For more information, we recommend studying references [1, 2, 3, 4, 5, 6, 7, 8, 9] and the citations within those sources. Mathematicians are studying fractional order differential equations due to their significant applications and realistic and practical models. Over the past ten years, numerous academics have examined the presence and distinctiveness of solutions to boundary value problems and associated interconnected systems for differential equations of fractional order (refer to [10, 11, 12, 13, 14, 15, 16, 17]). Consequently, the field dedicated to the study of existence theory has been extensively investigated. Nevertheless, the intricate structure of fractional derivatives makes it challenging to find analytical solutions for every fractional differential equation. Therefore, in such cases, it is most effective and beneficial to seek approximate solutions to these problems. Various numerical methods, including finite difference method, Fourier series method, Adomian decomposition method (ADM), inverse Laplace technique (ILT), variational iteration methods (VIM), fractional transform method (FTM), differential transform method (DTM), homotopy analysis method (HAM), method of radial base function (MRBM), wavelet techniques (WT), spectral methods, and others, have been developed to obtain numerical solutions for differential equations. For more details, refer to the cited sources [9, 18-38]. These methods have their own advantages and disadvantages. Some of them offer a really accurate estimation. Recently, several operational matrices have been developed to accomplish accurate approximation, as demonstrated in reference [39]. Subsequently, operational matrices were created for various wavelet techniques. This approach utilizes operational matrices, where each operation, such as differentiation and integration, involved in these equations is executed using a matrix. There is a wide range of operational matrices described in the literature for various orthogonal polynomials such as Legendre, Laguerre, Jacobi, and Bernstein polynomials [40, 41, 42, 43, 44, 45, 46, 47, 48]. In this chapter, we have created a numerical method using operational matrices with Bernstein polynomials, inspired by the aforementioned applications and uses of fractional differential equations. The argument we present is more comprehensive and does not require converting the vector of Bernstein polynomial functions to another basis, such as the block pulse function or Legendre polynomials. Based on our current understanding, the method we are considering offers a highly accurate estimation of the solution. Using these operational matrices, we implement our strategy to solve a single fractional order differential equation with specified boundary conditions as follows:

$$\begin{cases} D^{\alpha}y(t) + AD^{\mu}y(t) + By(t) = f(t), \quad 1 < \alpha \leq 2, \quad 0 < \mu \leq 1, \\ y(0) = a, \quad y(1) = b, \end{cases} \tag{1}$$

where $f(t)$ is the source term, A,B are any real numbers; then we extend our method to solve a boundary value problem of coupled system of fractional order differential equations of the form

$$\begin{cases} D^{\alpha}x(t) + A_1D^{\mu_1}x(t) + B_1D^{\nu_1}y(t) + C_1x(t) + D_1y(t) = f(t), \quad 1 < \alpha \leq 2, \quad 0 < \mu_1, \nu_1 \leq 1, \\ D^{\beta}y(t) + A_2D^{\mu_2}x(t) + B_2D^{\nu_2}y(t) + C_2x(t) + D_2y(t) = g(t), \quad 1 < \beta \leq 2, \quad 0 < \mu_2, \nu_2 \leq 1, \\ y(0) = a, \; y(1) = b, \; y(0) = c, \; y(1) = d, \end{cases} \tag{2}$$

where $f(t),g(t)$ are source terms of the system, A_i, B_i, C_i, D_i $(i=1,2)$ are any real constants. In addition, we assess the efficacy of our suggested technique by comparing our approximations to precise values and approximations obtained using other methods such as Jacobi polynomial approximations and Haar wavelets methods. We additionally offer illustrative examples to demonstrate our primary findings.

This chapter is structured into five sections. The initial part of the chapter includes references to fundamental works concerning the numerical and analytical solutions of differential equations of any order, employing diverse methodologies. Section 2 provides the essential concepts and outcomes of fractional calculus and Bernstein polynomials, as well as the development of certain operational matrices. In Section 3, we have presented the fundamental theory underlying the numerical technique. Section 4 includes compelling real-life illustrations and corresponding graphics. Section 5 provides a summary of the chapter.

2. BASIC DEFINITIONS AND RESULTS

In this section, we recall some fundamental definitions and results from the literature, which can be found in [10, 11, 12, 13, 14, 15, 16].

Definition 2.1. *The fractional integral of order $\gamma \in \mathbb{R}_+$ of a function $y \in L^1([0,1], \mathbb{R})$ is defined as*

$$I^{\gamma}_{0+}y(t) = \frac{1}{\Gamma(\gamma)}\int_0^t (t-\tau)^{\gamma-1}y(\tau)d\tau.$$

Definition 2.2. *The Caputo fractional order derivative of a function y on the interval* [0,1] *is defined by*

$$D^{\gamma}_{0+}y(t) = \frac{1}{\Gamma(n-\gamma)}\int_0^t (t-\tau)^{n-\gamma-1}y^{(n)}(\tau)d\tau,$$

where $n=[\gamma]+1$ and $[\gamma]$ represents the integer part of γ .

Lemma 2.1. *The fractional differential equation of order* $\gamma>0$

$$D^{\gamma}y(t) = 0, n-1<\gamma\leq n,$$

has a unique solution of the form $y(t)=d_0+d_1t+d_2t^2+\ldots+d_n-1t^{n-1}$, where $d_k \in R$ and $k=0,1, 2,3,\ldots, n-1$..

Lemma 2.2. *The following result holds for fractional differential equations*

$$I^{\gamma}D^{\gamma}y(t) = y(t) + d_0 + c_1t + d_2t^2 + \ldots + d_{n-1}t^{n-1},$$

for arbitrary $d_k \in R$, $k=0, 1,2,\ldots, n-1$.

Hence it follows that

$$D^{\gamma}t^k = \frac{\Gamma(k+1)}{\Gamma(k-\gamma+1)}t^{k-\gamma}, \quad I^{\gamma}t^k = \frac{\Gamma(k+1)}{\Gamma(k+\gamma+1)}t^{k+\gamma} \text{ and } D^{\gamma}[constant] = 0.$$

2.1 The Bernstein polynomials

The Bernstein polynomials $B_{i,m}(t)$ on [0,1] can be defined as

$$B_{i,m}(t) = \binom{m}{i}t^i(1-t)^{m-i}, \text{for } i = 0, 1, 2\ldots m,$$

where $\binom{m}{i} = \frac{m!}{(m-i)!i!}$, which on further simplification can be written in the most simplified form as

$$B_{i,m}(t) = \sum_{k=0}^{m-i} \Theta_{(i,k,m)} t^{k+i}, \quad i = 0, 1, 2 \dots m, \tag{3}$$

where

$$\Theta_{(i,k,m)} = (-1)^k \binom{m}{i} \binom{m-i}{k}.$$

Note that the sum of the Bernstein polynomials converges to 1.

Lemma 2.3. Convergence Analysis: *Assume that the function* $g \in C^{m+1}[0,1]$ *that is m+1 times continuously differentiable function and let* $X = \langle B_{0,m}, B_{1,m}, \dots, B_{m,m} \rangle$. *If* $C^T\Psi(x)$ *is the best approximation of g out of X, then the error bound is presented as*

$$\|g - C^T\Psi\|_2 \leq \frac{\sqrt{2} M S^{\frac{2m+3}{2}}}{\Gamma(m+2)\sqrt{2m+3}},$$

where $M = \max_{x \in [0,1]} |g^{(m+1)}(x)|, \quad S = \max\{1 - x_0, x_0\}$.

Proof. In view of Taylor polynomials, we have

$$F(x) = g(x_0) + (x - x_0)g^{(1)}(x_0) + \frac{(x-x_0)^2}{\Gamma 3} g^{(2)} + \dots + \frac{(x-x_0)^m}{\Gamma(m+1)} g^{(m)},$$

from which we know that

$$|g - F(x)| = |g^{(m+1)}(\eta)| \frac{(x-x_0)^{m+1}}{\Gamma(m+2)}, \quad \text{there exist } \eta \in (0,1).$$

Due the best approximation $C^T\Psi(x)$ of g, we have

$$\begin{aligned}
\|g - C^T\Psi(x)\|_2^2 &\leq \|g - F\|_2^2 \\
&= \int_0^1 (g(x) - F(x))^2 dx \\
&= \int_0^1 \left[|g^{(m+1)}(\eta)| \frac{(x-x_0)^{m+1}}{\Gamma(m+2)} \right]^2 dx \\
&\leq \frac{M^2}{\Gamma^2(m+2)} \int_0^1 (x-x_0)^{2m+2} dx \\
&\leq \frac{2M^2 S^{2m+3}}{\Gamma^2(m+2)(2m+3)}.
\end{aligned}$$

Hence we have

$$\|g - C^T\Psi(x)\|_2 \leq \frac{\sqrt{2}MS^{\frac{2m+3}{2}}}{\Gamma(m+2)\sqrt{(2m+3)}}. \quad \square$$

Let $H=L^2[0,1]$ be a Hilbert space, then the inner product can be defined as

$$<f,g> = \int_0^1 f(x).g(x)dx$$

and

$$Y = span\{B_{0,m}, B_{1,m}, \ldots B_{m,m}\}$$

is a finite dimensional and closed subspace. So if $f\in H$ is an arbitrary element then its best approximation is unique in Y. This terminology can be achieved by using $y_0\in Y$ and for all $y\in Y$, we have $\|f-y_0\|\leq\|f-y\|$. Thus any function can be approximated in terms of Bernstein polynomials as

$$f(t) = \sum_{i=0}^{m} c_i B_{(i,m)}, \tag{4}$$

where coefficient c_i can easily be calculated by multiplying (4) by $B_{(j,m)}(t)$, $j=0,1,2,\ldots m$ and integrating over [0,1] by using inner product and $d_i = \int_0^1 B_{(i,m)}(t)f(t)dt$, $\theta_{(i,j)} = \int_0^1 B_{(i,m)}(t)B_{(j,m)}(t)dt$, $i,j = 0,1,2.\ldots m$, we have

$$\int_0^1 f(t)B_{(j,m)}(t)dt = \int_0^1 \sum_{i=0}^{m} c_i B_{(i,m)}(t).B_{(j,m)}(t)dt, \quad j = 0,1,2\ldots m,$$

which implies that $\int_0^1 f(t)B_{(j,m)}(t)dt = \sum_{i=0}^{m} c_i \int_0^1 B_{(i,m)}(t).B_{(j,m)}(t)dt, \quad j = 0,1,2\ldots m$

$$\text{which implies that } [d_0\ d_1\ \ldots.\ d_m] = [c_0\ c_1.\ldots c_m]\begin{bmatrix} \theta_{(0,0)} & \theta_{(0,1)} & \cdots & \theta_{(0,r)} & \cdots & \theta_{(0,m)} \\ \theta_{(1,0)} & \theta_{(1,1)} & \cdots & \theta_{(1,r)} & \cdots & \theta_{(1,m)} \\ \vdots & \vdots & \vdots & \vdots & \vdots & \vdots \\ \theta_{(r,0)} & \theta_{(r,1)} & \cdots & \theta_{(r,r)} & \cdots & \theta_{(r,m)} \\ \vdots & \vdots & \vdots & \vdots & \vdots & \vdots \\ \theta_{(m,0)} & \theta_{(m,1)} & \cdots & \theta_{(m,r)} & \cdots & \theta_{(m,m)} \end{bmatrix}. \tag{5}$$

Let $X_M=[d_0 d_1....d_m]$, $C_M=[c_0 c_1....c_m]$, where $M=m+1$ where M is the scale level and $\Phi_{M\times M} = \begin{bmatrix} \theta_{(0,0)} & \theta_{(0,1)} & \cdots & \theta_{(0,r)} & \cdots & \theta_{(0,m)} \\ \theta_{(1,0)} & \theta_{(1,1)} & \cdots & \theta_{(1,r)} & \cdots & \theta_{(1,m)} \\ \vdots & \vdots & \vdots & \vdots & \vdots & \vdots \\ \theta_{(r,0)} & \theta_{(r,1)} & \cdots & \theta_{(r,r)} & \cdots & \theta_{(r,m)} \\ \vdots & \vdots & \vdots & \vdots & \vdots & \vdots \\ \theta_{(m,0)} & \theta_{(m,1)} & \cdots & \theta_{(m,r)} & \cdots & \theta_{(m,m)} \end{bmatrix}$, so

$$X_M = C_M \Phi_{M\times M} \Rightarrow C_M = X_M \Phi^{-1}_{M\times M}. \tag{6}$$

where $\Phi_{m\times M}$ is called the dual matrix of the Bernstein polynomials. After calculating $ci_,$ (4) can be written as

$$f(t) = C_M B_M^T(t), \quad C_M \text{ is coefficient matrix}$$

where

$$B_M(t) = \left[B_{(0,m)}, B_{(1,m)}, \ldots . B_{(m,m)}\right]. \tag{7}$$

Lemma 2.4. *Let $B^T_M(t)$ be the function vector defined in* (3), *then the fractional order integration of $B^T_M(t)$ is given by*

$$I^\alpha B_M^T(t) = P^\alpha_{M\times M} B_M^T(t), \tag{8}$$

where P^α_{MXM} is the fractional integration's operational matrix defined as

$$P^\alpha_{M\times M} = \hat{P}^\alpha_{M\times M} \Phi^{-1}_{M\times M}$$

and $\Phi^{-1}_{M\times M}$ is given in (3) and $P^\alpha_{M\times M}$ is given by

$$\hat{P}^\alpha_{M\times M} = \begin{bmatrix} \Psi_{(0,0)} & \Psi_{(0,1)} & \cdots & \Psi_{(0,r)} & \cdots & \Psi_{(0,m)} \\ \Psi_{(1,0)} & \Psi_{(1,1)} & \cdots & \Psi_{(1,r)} & \cdots & \Psi_{(1,m)} \\ \vdots & \vdots & \vdots & \vdots & \vdots & \vdots \\ \Psi_{(r,0)} & \Psi_{(r,1)} & \cdots & \Psi_{(r,r)} & \cdots & \Psi_{(r,m)} \\ \vdots & \vdots & \vdots & \vdots & \vdots & \vdots \\ \Psi_{(m,0)} & \Psi_{(m,1)} & \cdots & \Psi_{(m,r)} & \cdots & \Psi_{(m,m)} \end{bmatrix}, \tag{9}$$

where

$$\Psi_{i,j} = \sum_{k=0}^{m-i} \sum_{l=0}^{m-j} \theta_{(i,k,m)} \theta_{(j,l,m)} \frac{\Gamma(k+i+1)}{(i+j+k+l+\alpha+1)\Gamma(k+i+\alpha+1)}. \tag{10}$$

Proof. Consider

$$B_{i,m}(t) = \sum_{k=0}^{m-i} \theta_{(i,k,m)} t^{k+i} \tag{11}$$

taking fractional integration of both sides, we have

$$I^{\alpha} B_{i,m}(t) = \sum_{k=0}^{m-i} \theta_{(i,k,m)} I^{\alpha} t^{k+i} = \sum_{k=0}^{m-i} \theta_{(i,k,m)} \frac{\Gamma(k+i+\alpha)}{\Gamma(k+i+\alpha+1)} t^{k+i+\alpha}. \tag{12}$$

Now to approximate right-hand sides of above

$$\sum_{k=0}^{m-i} \theta_{(i,k,m)} \frac{\Gamma(k+i+\alpha)}{\Gamma(k+i+\alpha+1)} t^{k+i+\alpha} = C_M^{(i)} B_M^T(t) \tag{13}$$

where $C^{(i)}_M$ can be approximated by using (3) as

$$C_M^{(i)} = X_M^{(i)} \Phi_{M\times M}^{-1}, \tag{14}$$

where entries of the vector $X^{(i)}_M$ can be calculated in generalized form as

$$X_M^{(j)} = \int_0^1 \sum_{k=0}^{m-i} \theta_{(i,k,m)} \frac{\Gamma(k+i+\alpha)}{\Gamma(k+i+\alpha+1)} t^{k+i+\alpha} B_{j,m}(t) dt, \quad j = 0,1,2. \dots m$$

$$\Rightarrow X_M^{(j)} = \int_0^1 \sum_{k=0}^{m-i} \theta_{(i,k,m)} \frac{\Gamma(k+i+\alpha)}{\Gamma(k+i+\alpha+1)} \sum_{l=0}^{m-j} \theta_{(j,l,m)} t^{k+i+\alpha} . t^{l+j} dt, \quad j = 0,1,2. \dots m$$

$$= \sum_{k=0}^{m-i} \theta_{(i,k,m)} \sum_{l=0}^{m-j} \theta_{(j,l,m)} \frac{\Gamma(k+i+\alpha)}{\Gamma(k+i+\alpha+1)} \frac{1}{(k+l+j+i+\alpha+1)}, \quad j = 0,1,2,\dots m \tag{15}$$

evaluating this result for i = 0,1,2....m, we have

$$\begin{bmatrix} I^{\alpha} B_{0,m}(t) \\ I^{\alpha} B_{1,m}(t) \\ \vdots \\ \vdots \\ \vdots \\ I^{\alpha} B_{m,m}(t) \end{bmatrix} = \begin{bmatrix} X_M^{(0)} \Phi_{M\times M}^{-1} B_M^T(t) \\ X_M^{(1)} \Phi_{M\times M}^{-1} B_M^T(t) \\ \vdots \\ \vdots \\ \vdots \\ X_M^{(m)} \Phi_{M\times M}^{-1} B_M^T(t)) \end{bmatrix} \tag{16}$$

further writing

$$\Psi_{i,j} = \sum_{k=0}^{m-i} \sum_{l=0}^{m-j} \theta_{(i,k,m)} . \theta_{(j,l,m)} \frac{\Gamma(k+i+\alpha)}{\Gamma(k+i+\alpha+1)} \frac{1}{(k+l+j+i+\alpha+1)}$$

we get

$$I^{\alpha}B_M^T(t) = \hat{P}^{\alpha}_{M\times M}.\Phi^{-1}_{M\times M}.B_M^T(t). \tag{17}$$

Let us represent

$$\hat{P}^{\alpha}_{M\times M}.\Phi^{-1}_{M\times M} = P^{\alpha}_{M\times M}$$

thus

$$I^{\alpha}B_M^T(t) = P^{\alpha}_{M\times M}.B_M^T(t). \tag{18}$$

Lemma 2.5. *Let $B^T_M(t)$ be the function vector as defined in (3), then fractional order derivative is defined as*

$$D^{\alpha}B_M^T(t) = G^{\alpha}_{M\times M}.B_M^T(t) \tag{19}$$

where $G^{\alpha}_{M\times M}$ is the operational matrix of fractional order derivative given by

$$G^{\alpha}_{M\times M} = \hat{G}^{\alpha}_{M\times M}\Phi^{-1}_{M\times M}, \tag{20}$$

where $\Phi_{M\times M}$ is the dual matrix given in (3) *and*

$$\hat{G}^{\alpha}_{M\times M} = \begin{bmatrix} \Psi_{(0,0)} & \Psi_{(0,1)} & \cdots & \Psi_{(0,r)} & \cdots & \Psi_{(0,m)} \\ \Psi_{(1,0)} & \Psi_{(1,1)} & \cdots & \Psi_{(1,r)} & \cdots & \Psi_{(1,m)} \\ \vdots & \vdots & \vdots & \vdots & \vdots & \vdots \\ \Psi_{(r,0)} & \Psi_{(r,1)} & \cdots & \Psi_{(r,r)} & \cdots & \Psi_{(r,m)} \\ \vdots & \vdots & \vdots & \vdots & \vdots & \vdots \\ \Psi_{(m,0)} & \Psi_{(m,1)} & \cdots & \Psi_{(m,r)} & \cdots & \Psi_{(m,m)} \end{bmatrix}, \tag{21}$$

where $\Psi_{(i,j)}$ is defined for two different cases as

Case I: ($i<\lceil\alpha\rceil$)

$$\Psi_{i,j} = \sum_{k=\lceil\alpha\rceil}^{m-i}\sum_{l=0}^{m-j}\theta_{(i,k,m)}.\theta_{(j,l,m)}\frac{\Gamma(k+i-\alpha)}{\Gamma(k+i-\alpha+1)}\frac{1}{(k+l+j+i-\alpha+1)} \tag{22}$$

Case II: ($i\geq\lceil\alpha\rceil$)

$$\Psi_{i,j} = \sum_{k=0}^{m-i}\sum_{l=0}^{m-j}\theta_{(i,k,m)}.\theta_{(j,l,m)}\frac{\Gamma(k+i-\alpha)}{\Gamma(k+i-\alpha+1)}\frac{1}{(k+l+j+i-\alpha+1)}. \tag{23}$$

Proof. Consider the general element as

$$D^{\alpha}B_{i,m}(t) = D^{\alpha}\left(\sum_{k=0}^{m-i}\theta_{(i,k,m)}.t^{k+i}\right) = \sum_{k=0}^{m-i}\theta_{(i,k,m)}D^{\alpha}t^{k+i}. \tag{24}$$

It is to be noted in the polynomial function $B_{i,m}$ the power of the variable 't' is an ascending order and the lowest power is 'i' therefore the first [α−1] terms becomes zero when we take derivative of order α.

Case I: (i<[α]) By the use of definition of fractional derivative

$$D^{\alpha}B_{i,m}(t) = \sum_{k=[\alpha]}^{m-i} \theta_{(i,k,m)} \frac{\Gamma(k+i+1)}{\Gamma(k+i-\alpha+1)} t^{k+i-\alpha}. \tag{25}$$

Now approximating RHS of (25) as

$$\sum_{k=[\alpha]}^{m-i} \theta_{(i,k,m)} \frac{\Gamma(k+i+1)}{\Gamma(k+i-\alpha+1)} t^{k+i-\alpha} = C_M^{(i)} B_M^T(t) \tag{26}$$

further implies that

$$X_M^{(j)} = \int_0^1 \sum_{k=[\alpha]}^{m-i} \theta_{(i,k,m)} \frac{\Gamma(k+i+1)}{\Gamma(k+i-\alpha+1)} t^{k+i-\alpha} B_{j,m}(t)dt, \quad j = 0,1,2.\dots m$$

$$\Rightarrow X_M^{(j)} = \sum_{k=[\alpha]}^{m-i} \theta_{(i,k,m)} \sum_{l=0}^{m-j} \theta_{(j,l,m)} \frac{\Gamma(k+i+1)}{\Gamma(k+i-\alpha+1)(k+i+l+j-\alpha+1)}, \quad j = 0,1,2.\dots m \tag{27}$$

Case II: (i≥[α]) if i≤[α] then

$$X_M^{(j)} = \sum_{k=0}^{m-i} \theta_{(i,k,m)} \sum_{l=0}^{m-j} \theta_{(j,l,m)} \frac{\Gamma(k+i+1)}{\Gamma(k+i-\alpha+1)(k+i+l+j-\alpha+1)}, \quad j = 0,1,2.\dots m. \tag{28}$$

After careful simplification, we get

$$\begin{bmatrix} D^{\alpha}B_{0,m}(t) \\ D^{\alpha}B_{1,m}(t) \\ \vdots \\ \vdots \\ \vdots \\ D^{\alpha}B_{m,m}(t) \end{bmatrix} = \begin{bmatrix} X_M^{(0)}\Phi_{M\times M}^{-1}B_M^T(t) \\ X_M^{(1)}\Phi_{M\times M}^{-1}B_M^T(t) \\ \vdots \\ \vdots \\ \vdots \\ X_M^{(m)}\Phi_{M\times M}^{-1}B_M^T(t)) \end{bmatrix}. \tag{29}$$

On further simplification, we have

$$\Psi_{i,j} = \sum_{k=[\alpha]}^{m-i}\sum_{l=0}^{m-j} \theta_{(i,k,m)}.\theta_{(j,l,m)} \frac{\Gamma(k+i+1)}{\Gamma(k+i-\alpha+1)} \frac{1}{(k+l+j+i-\alpha+1)} \quad (i<[\alpha])$$

$$\Psi_{i,j} = \sum_{k=0}^{m-i}\sum_{l=0}^{m-j} \theta_{(i,k,m)}.\theta_{(j,l,m)} \frac{\Gamma(k+i+1)}{\Gamma(k+i-\alpha+1)} \frac{1}{(k+l+j+i-\alpha+1)}$$

we get $D^{\alpha}B_M^T(t) = \hat{G}^{\alpha}_{M\times M}\Phi^{-1}_{M\times M}.B_M^T(t)$.

(30)

Let

$$\hat{G}^{\alpha}_{M\times M}\Phi^{-1}_{M\times M} = G^{\alpha}_{M\times M}$$

so

$$D^{\alpha}B_M^T(t) = G^{\alpha}_{M\times M}B_M^T(t)$$

which is the desired result.

Lemma 2.6. *An operational matrix corresponding to the boundary condition by taking $B^T{}_M$ (t) is function vector and K is coefficient vector by taking the approximation*

$$u(t) = K\hat{B}(t)$$

is given by

$$Q^{\alpha,\phi}_{M\times M} = \begin{bmatrix} \Omega_{(0,0)} & \Omega_{(0,1)} & \cdots & \Omega_{(0,r)} & \cdots & \Omega_{(0,m)} \\ \Omega_{(1,0)} & \Omega_{(1,1)} & \cdots & \Omega_{(1,r)} & \cdots & \Omega_{(1,m)} \\ \vdots & \vdots & \vdots & \vdots & \vdots & \vdots \\ \Omega_{(r,0)} & \Omega_{(r,1)} & \cdots & \Omega_{(r,r)} & \cdots & \Omega_{(r,m)} \\ \vdots & \vdots & \vdots & \vdots & \vdots & \vdots \\ \Omega_{(m,0)} & \Omega_{(m,1)} & \cdots & \Omega_{(m,r)} & \cdots & \Omega_{(m,m)} \end{bmatrix}, \quad (31)$$

where

$$\Omega_{i,j} = \int_0^1 \Delta_{i,m}\phi(t)B_j(t)dt, \quad i,j = 0,1,2. \ldots m.$$

Proof. Let us take $u(t) = K\hat{B}(t)$, then

$$ {}_0I_1^{\alpha}K\hat{B}(t) = K{}_0I_1^{\alpha}\hat{B}(t) = K\begin{bmatrix} {}_0I_1^{\alpha}B_0(t) \\ {}_0I_1^{\alpha}B_1(t) \\ \vdots \\ \vdots \\ \vdots \\ {}_0I_1^{\alpha}B_m(t) \end{bmatrix} $$

Let us evaluate the general terms

$$ \begin{aligned} {}_0I_1^{\alpha}B_i(t)dt &= \frac{1}{\Gamma\alpha}\int_0^1 (1-s)^{\alpha-1}B_{i,m}(s)ds \\ &= \frac{1}{\Gamma\alpha}\sum_{k=0}^{m-i}\Theta_{i,k,m}\int_0^1 (1-s)^{\alpha-1}s^{k+i}ds. \end{aligned} \tag{32} $$

Since by

$$ L\left(\int_0^1 (1-s)^{\alpha-1}s^{k+i}ds\right) = \frac{\Gamma\alpha\Gamma(k+i+1)}{\tau^{k+\alpha+i}} $$

taking inverse Laplace of both sides, we get

$$ \int_0^1 (1-s)^{\alpha-1}.s^{k+i}ds = L^{-1}\left[\frac{\Gamma\alpha.\Gamma(k+i+1)}{\tau^{k+\alpha+i}}\right] = \frac{\Gamma\alpha.\Gamma(k+i+1)}{\Gamma(k+i+\alpha+1)} $$

now Eq. (32) implies that

$$ {}_0I_1^{\alpha}B_i(t)dt = \sum_{k=0}^{m-i}\Theta_{i,k,m}\frac{\Gamma(k+i+1)}{\Gamma(k+i+\alpha+1)} = \Delta_{i,m}. \tag{33} $$

Now using the approximation $\Delta_{i,m}\phi(t) = \sum_{i=0}^{m}\hat{c}_iB_i(t) = C_M^iB_M^T$, and using Eq. (3) we get $C_M^i = K_M^i\Phi_{M\times M}^{-1}B_M^T$ and using $c_j = \int_0^1\phi(t)B_j(t)dt$,

$$\phi(t)KI^{\alpha}\hat{B}(t) = K\begin{bmatrix} \Delta_{0,m}\phi(t) \\ \Delta_{1,m}\phi(t) \\ \vdots \\ \vdots \\ \vdots \\ \Delta_{m,m}\phi(t) \end{bmatrix} = K\begin{bmatrix} C_M^0\Phi_{M\times M}^{-1}B_M^T(t) \\ C_M^1\Phi_{M\times M}^{-1}B_M^T(t) \\ \vdots \\ \vdots \\ \vdots \\ C_M^m\Phi_{M\times M}^{-1}B_M^T(t) \end{bmatrix}$$

$$= K\begin{bmatrix} c_0^0 & c_1^0 & \cdots & c_r^0 & \cdots & c_m^0 \\ c_0^1 & c_1^1 & \cdots & c_r^1 & \cdots & c_m^1 \\ \vdots & \vdots & \vdots & \vdots & \vdots & \vdots \\ c_0^r & c_1^r & \cdots & c_r^r & \cdots & c_m^r \\ \vdots & \vdots & \vdots & \vdots & \vdots & \vdots \\ c_0^m & c_1^m & \cdots & c_r^m & \cdots & c_m^m \end{bmatrix}\begin{bmatrix} \Phi_{M\times M}^{-1}B_M^T(t) \\ \Phi_{M\times M}^{-1}B_M^T(t) \\ \vdots \\ \vdots \\ \vdots \\ \Phi_{M\times M}^{-1}B_M^T(t) \end{bmatrix}. \tag{34}$$

On further simplification, we get

$$\phi(t)KI^{\alpha}\hat{B}(t) = K\begin{bmatrix} \Omega_{(0,0)} & \Omega_{(0,1)} & \cdots & \Omega_{(0,r)} & \cdots & \Omega_{(0,m)} \\ \Omega_{(1,0)} & \Omega_{(1,1)} & \cdots & \Omega_{(1,r)} & \cdots & \Omega_{(1,m)} \\ \vdots & \vdots & \vdots & \vdots & \vdots & \vdots \\ \Omega_{(r,0)} & \Omega_{(r,1)} & \cdots & \Omega_{(r,r)} & \cdots & \Omega_{(r,m)} \\ \vdots & \vdots & \vdots & \vdots & \vdots & \vdots \\ \Omega_{(m,0)} & \Omega_{(m,1)} & \cdots & \Omega_{(m,r)} & \cdots & \Omega_{(m,m)} \end{bmatrix}\begin{bmatrix} B_0(t) \\ B_1(t) \\ \vdots \\ \vdots \\ \vdots \\ B_m(t) \end{bmatrix}. \tag{35}$$

So

$$\phi(t)\,{}_0I_1^{\alpha}u(t) = KQ_{M\times M}^{\alpha,\phi}B_M^T(t),$$

and

$$\Omega_{i,j} = \int_0^1 \Delta_{i,m}\phi(t)B_j(t)dt, \quad i,j = 0,1,2.\dots m. \tag{36}$$

which is the required result.

3. APPLICATIONS OF OPERATIONAL MATRICES

In this section, we will be approximating a boundary value problem using a differential equation of fractional order, as well as a linked system of boundary value problems with fractional order. The utilization of acquired operational matrices is demonstrated in the subsequent step.

3.1 Fractional differential equations

Consider the following problem in generalized form of fractional order differential equation

$$\begin{aligned} &D^{\alpha}y(t)+AD^{\mu}y(t)+By(t)=f(t), \quad 1<\alpha\leq 2, \quad 0<\mu\leq 1, \\ &\text{subject to the boundary conditions } y(0)=a, \quad y(1)=b, \end{aligned} \tag{37}$$

where $f(t)$ is a source term; A,B are any real constants and $y(t)$ is an unknown solution which we want to determine. To obtain a numerical solution of the above problem in terms of Bernstein polynomials, we proceed as

$$\text{Let } D^{\alpha}y(t)=K_M B_M^T(t). \tag{38}$$

Applying fractional integral of order α we have

$$y(t)=K_M P_{M\times M}^{\alpha}B_M^T(t)-c_0+c_1t$$

using boundary conditions, we have

$$c_0=a, \quad c_1=b-a-K_M P_{M\times M}^{\alpha}B_M^T(t)\big|_{t=1}.$$

Using the approximation and Lemma 2.2

$$a+t(b-a)=F_M^{(1)}B_M^T(t), \quad tP_{M\times M}^{\alpha}B_M^T(t)\Big|_{t=1}=Q_{M\times M}^{\alpha,\phi}B_M^T(t).$$

Hence

$$y(t)=K_M P_{M\times M}^{\alpha}B_M^T(t)+a+t(b-a)-tK_M P_{M\times M}^{\alpha}B_M^T(t)\big|_{t=1},$$

$$\begin{aligned} \text{which gives } y(t)&=K_M P_{M\times M}^{\alpha}B_M^T(t)+F_M^{(1)}B_M^T(t)-Q_{M\times M}^{\alpha,\phi}B_M^T(t) \\ &=K_M\left(P_{M\times M}^{\alpha}-Q_{M\times M}^{\alpha,\phi}\right)B_M^T(t)+F_M^{(1)}B_M^T(t). \end{aligned} \tag{39}$$

Now

$$\begin{aligned} D^{\mu}y(t)&=D^{\mu}\left[K_M\left(P_{M\times M}^{\alpha}-Q_{M\times M}^{\alpha,\phi}\right)B_M^T(t)+F_M^{(1)}B_M^T(t)\right] \\ &=K_M\left(P_{M\times M}^{\alpha}-Q_{M\times M}^{\alpha,\phi}\right)G_{M\times M}^{\mu}B_M^T(t)+F_M^{(1)}G_{M\times M}^{\mu}B_M^T(t) \end{aligned} \tag{40}$$

and

$$f(t)=F_M^{(2)}B_M^T(t). \tag{41}$$

Putting Eqs. (38)–(41) in Eq. (37), we get

$$\begin{aligned} &K_M B_M^T(t)+AK_M\left(P_{M\times M}^{\alpha}-Q_{M\times M}^{\alpha,\phi}\right)G_{M\times M}^{\mu}B_M^T(t)+AF_M^{(1)}G_{M\times M}^{\mu}B_M^T(t) \\ &\quad+BK_M\left(P_{M\times M}^{\alpha}-Q_{M\times M}^{\alpha,\phi}\right)B_M^T(t)+BF_M^{(1)}B_M^T(t)=F_M^{(2)}B_M^T(t). \end{aligned} \tag{42}$$

In simple form, we can write (42) as

$$K_M B_M^T(t) + AK_M\left(P_{M\times M}^{\alpha} - Q_{M\times M}^{\alpha,\phi}\right)G_{M\times M}^{\mu}B_M^T(t) + AF_M^{(1)}G_{M\times M}^{\mu}B_M^T(t)$$

$$+ BK_M\left(P_{M\times M}^{\alpha} - Q_{M\times M}^{\alpha,\phi}\right)B_M^T(t) + BF_M^{(1)}B_M^T(t) - F_M^{(2)}B_M^T(t) = 0 \qquad (43)$$

$$K_M + K_M\left(A\hat{P}_{M\times M}^{\alpha}G_{M\times M}^{\mu} + B\hat{P}_{M\times M}^{\alpha}\right) + AF_M^{(1)}G_{M\times M}^{\mu} + BF_M^{(1)} - F_M^{(2)},$$

where

$$\hat{P}_{M\times M}^{\alpha} = P_{M\times M}^{\alpha} - Q_{M\times M}^{\alpha,\phi}.$$

Eq. (43) is an algebraic equation of Lyapunov type, which can be easily solved for the unknown coefficient vector K_M. When we find K_M, then putting this in Eq. (39), we get the required approximate solution of the problem.

3.2 Coupled system of boundary value problem of fractional order differential equations

Consider a coupled system of a fractional order boundary value problem

$$D^{\alpha}x(t) + A_1D^{\mu_1}x(t) + B_1D^{\nu_1}y(t) + C_1x(t) + D_1y(t) = f(t), \quad 1<\alpha\le 2, 0<\mu_1,\nu_1\le 1,$$
$$D^{\beta}y(t) + A_2D^{\mu_2}x(t) + B_2D^{\nu_2}y(t) + C_2x(t) + D_2y(t) = g(t), \quad 1<\beta\le 2, 0<\mu_2,\nu_2\le 1, \qquad (44)$$

subject to the boundary conditions

$$x(0) = a, \quad x(1) = b \quad y(0) = c, \quad y(1) = d, \qquad (45)$$

where $A_i,B_i,C_i,D_i (i=1,2)$ are any real constants, $f(t), g(t)$ are given source terms. We approximate the solution of the above system in terms of Bernstein polynomials such as

$$D^{\alpha}x(t) = K_M B_M^T(t), \quad D^{\beta}y(t) = L_M B_M^T(t)$$

$$x(t) = K_M P_{M\times M}^{\alpha}B_M^T(t) + c_0 + c_1t, \quad y(t) = L_M\left(P_{M\times M}^{\beta}B_M^T(t) + d_0 + d_1t\right.$$

applying boundary conditions, we have

$$x(t) = K_M(P_{M\times M}^{\alpha}B_M^T(t) + a + t(b-a) - tK_M P_{M\times M}^{\alpha}B_M^T(t)|_{t=1},$$

$$y(t) = K_M(P_{M\times M}^{\beta}B_M^T(t) + c + t(d-c) - tK_M P_{M\times M}^{\beta}B_M^T(t)\Big|_{t=1}.$$

let us approximate

$$a + t(b-a) = F_M^1 B_M^T(t), c + t(d-c) = F_M^2 B_M^T(t)$$

$$tP_{M\times M}^{\alpha}B_M^T(t)\Big|_{t=1} = Q_{M\times M}^{\alpha,\phi}B_M^T(t), tP_{M\times M}^{\beta}B_M^T(t)\Big|_{t=1} = Q_{M\times M}^{\beta,\phi}B_M^T(t)$$

Then

$$x(t) = K_M P^{\alpha}_{M\times M} B^T_M(t) + F^{(1)}_M B^T_M(t) - K_M Q^{\alpha,\phi}_{M\times M} B^T_M(t)$$

$$y(t) = L_M P^{\beta}_{M\times M} B^T_M(t) + F^{(2)}_M B^T_M(t) - L_M Q^{\beta,\phi}_{M\times M} B^T_M(t)$$

$$D^{\mu_1} x(t) = \left[K_M P^{\alpha}_{M\times M} B^T_M(t) + F^{(1)}_M B^T_M(t) - K_M Q^{\alpha,\phi}_{M\times M} B^T_M(t)\right]$$

$$= K_M \left(P^{\alpha}_{M\times M} - Q^{\alpha,\phi}_{M\times M}\right) G^{\mu_1}_{M\times M} + F^{(1)}_M G^{\mu_1}_{M\times M} B^T_M(t)$$

$$D^{\nu_1} y(t) = D^{\nu_1}\left[L_M P^{\beta}_{M\times M} B^T_M(t) + F^{(2)}_M B^T_M(t) - L_M Q^{\beta,\phi}_{M\times M} B^T_M(t)\right]$$

$$= L_M \left(P^{\beta}_{M\times M} - Q^{\beta,\phi}_{M\times M}\right) G^{\nu_1}_{M\times M} + F^{(2)}_M G^{\nu_1}_{M\times M} B^T_M(t)$$

$$D^{\mu_2} x(t) = D^{\mu_2}\left[K_M P^{\alpha}_{M\times M} B^T_M(t) + F^{(1)}_M B^T_M(t) - K_M Q^{\alpha,\phi}_{M\times M} B^T_M(t)\right]$$

$$= K_M \left(P^{\alpha}_{M\times M} - Q^{\alpha,\phi}_{M\times M}\right) G^{\mu_2}_{M\times M} + F^{(1)}_M G^{\mu_2}_{M\times M} B^T_M(t)$$

and

$$D^{\nu_2} y(t) = D^{\nu_2}\left[K_M P^{\beta}_{M\times M} B^T_M(t) + F^{(2)}_M B^T_M(t) - K_M Q^{\beta,\phi}_{M\times M} B^T_M(t)\right]$$

$$= L_M \left(P^{\beta}_{M\times M} - Q^{\beta,\phi}_{M\times M}\right) G^{\nu_2}_{M\times M} + F^{(2)}_M G^{\nu_2}_{M\times M} B^T_M(t)$$

$$f(t) = F^{(3)} B^T_M(t), \quad g(t) = F^{(4)} B^T_M(t).$$

Thus system (44) implies that

$$\begin{aligned}
&K_M B^T_M(t) + A_1 K_M \left(P^{\alpha}_{M\times M} - Q^{\alpha,\phi}_{M\times M}\right) G^{\mu_1}_{M\times M} + A_1 F^{(1)}_M G^{\mu_1}_{M\times M} B^T_M(t)\\
&\quad + B_1 L_M \left(P^{\beta}_{M\times M} - Q^{\beta,\phi}_{M\times M}\right) G^{\nu_1}_{M\times M} + B_1 F^{(2)}_M G^{\nu_1}_{M\times M} B^T_M(t) + C_1 K_M P^{\alpha}_{M\times M} B^T_M(t)\\
&\quad + C_1 F^{(1)}_M B^T_M(t) - C_1 K_M Q^{\alpha,\phi}_{M\times M} B^T_M(t) + D_1 L_M P^{\beta}_{M\times M} B^T_M(t) + D_1 F^{(2)}_M B^T_M(t)\\
&\quad - D_1 L_M Q^{\beta,\phi}_{M\times M} B^T_M(t) = F^{(3)} B^T_M(t)
\end{aligned}$$

$$L_M B_M^T(t) + A_2 K_M \left(P_{M\times M}^{\alpha} - Q_{M\times M}^{\alpha,\phi} \right) G_{M\times M}^{\mu_2} + A_2 F_M^{(1)} G_{M\times M}^{\mu_2} B_M^T(t)$$
$$+ B_2 L_M \left(P_{M\times M}^{\beta} - Q_{M\times M}^{\beta,\phi} \right) G_{M\times M}^{\upsilon_2} + B_2 F_M^{(2)} G_{M\times M}^{\upsilon_2} B_M^T(t) + C_2 K_M P_{M\times M}^{\alpha} B_M^T(t)$$
$$+ C_2 F_M^{(1)} B_M^T(t) - C_2 K_M Q_{M\times M}^{\alpha,\phi} B_M^T(t) + D_2 L_M P_{M\times M}^{\beta} B_M^T(t) + D_2 F_M^{(2)} B_M^T(t)$$
$$- D_2 L_M Q_{M\times M}^{\beta,\phi} B_M^T(t) = F^{(4)} B_M^T(t). \tag{46}$$

Rearranging the terms in the above system and using the following notation for simplicity in Eq.(46)

$$\hat{Q}_{M\times M}^{\alpha} = A_1 \left(P_{M\times M}^{\alpha} - Q_{M\times M}^{\alpha,\phi} \right) G_{M\times M}^{\mu_1} + C_1 \left(P_{M\times M}^{\alpha} - Q_{M\times M}^{\alpha,\phi} \right)$$

$$\hat{Q}_{M\times M}^{\beta} = B_1 \left(P_{M\times M}^{\beta} - Q_{M\times M}^{\beta,\phi} \right) G_{M\times M}^{\upsilon_1} + D_1 \left(P_{M\times M}^{\beta} - Q_{M\times M}^{\beta,\phi} \right)$$

$$\hat{R}_{M\times M}^{\alpha} = A_2 \left(P_{M\times M}^{\alpha} - Q_{M\times M}^{\alpha,\phi} \right) G_{M\times M}^{\mu_2} + C_2 \left(P_{M\times M}^{\alpha} - Q_{M\times M}^{\alpha,\phi} \right)$$

$$\hat{R}_{M\times M}^{\beta} = B_2 \left(P_{M\times M}^{\beta} - Q_{M\times M}^{\beta,\phi} \right) G_{M\times M}^{\upsilon_2} + D_2 \left(P_{M\times M}^{\beta} - Q_{M\times M}^{\beta,\phi} \right)$$

$$F_M = A_1 F_M^{(1)} G_{M\times M}^{\mu_1} + B_1 F_M^{(2)} G_{M\times M}^{\upsilon_1} + C_1 F_M^{(1)} + F_M^{(2)} - D_1 F_M^{(3)}$$

$$G_M = A_2 F_M^{(1)} G_{M\times M}^{\mu_2} + B_2 F_M^{(2)} G_{M\times M}^{\upsilon_2} + C_2 F_M^{(1)} + D_2 F_M^{(2)} - F_M^{(4)},$$

the above system (46) becomes

$$K_M B_M^T(t) + K_M \hat{Q}_{M\times M}^{\alpha} B_M^T(t) + L_M \hat{Q}_{M\times M}^{\beta} B_M^T(t) + F_M B_M^T(t) = 0$$

$$L_M B_M^T(t) + K_M \hat{R}_{M\times M}^{\alpha} B_M^T(t) + L_M \hat{R}_{M\times M}^{\beta} B_M^T(t) + G_M B_M^T(t) = 0$$

$$[K_M \;\; L_M] \begin{bmatrix} B_M^T(t) & 0 \\ 0 & B_M^T(t) \end{bmatrix} + [K_M \;\; L_M] \begin{bmatrix} \hat{Q}_{M\times M}^{\alpha} & 0 \\ 0 & \hat{R}_{M\times M}^{\beta} \end{bmatrix} \begin{bmatrix} B_M^T(t) & 0 \\ 0 & B_M^T(t) \end{bmatrix}$$

$$+ [K_M \;\; L_M] \begin{bmatrix} 0 & \hat{R}_{M\times M}^{\alpha} \\ \hat{Q}_{M\times M}^{\beta} & 0 \end{bmatrix} \begin{bmatrix} B_M^T(t) & 0 \\ 0 & B_M^T(t) \end{bmatrix} + [F_M \;\; G_M] \begin{bmatrix} B_M^T(t) & 0 \\ 0 & B_M^T(t) \end{bmatrix} = 0$$

$$[K_M \;\; L_M] + [K_M \;\; L_M] \begin{bmatrix} \hat{Q}_{M\times M}^{\alpha} & \hat{R}_{M\times M}^{\alpha} \\ \hat{Q}_{M\times M}^{\beta} & \hat{R}_{M\times M}^{\beta} \end{bmatrix} + [F_M \;\; G_M] = 0, \tag{47}$$

which is an algebraic equation that can be easily solved by using Matlab functional solver or Mathematica for unknown matrix $[K_M \, L_M]$. Calculating the coefficient matrix K_M, L_M and putting it in equations

$$x(t) = K_M P^{\alpha}_{M\times M} B^T_M(t) + F^{(1)}_M B^T_M(t) - K_M Q^{\alpha,\phi}_{M\times M} B^T_M(t)$$

$$y(t) = L_M P^{\beta}_{M\times M} B^T_M(t) + F^{(2)}_M B^T_M(t) - L_M Q^{\beta,\phi}_{M\times M} B^T_M(t),$$

we get the required approximate solution.

4. APPLICATIONS OF OUR METHOD TO SOME EXAMPLES

Example 4.1. *Consider*

$$D^{\alpha}y(t) + c_1 D^{\nu}y(t) + c_2 y(t) = f(t), \quad 1<\alpha<2 \tag{48}$$

subject to the boundary conditions

$$y(0) = 0, \quad y(1) = 0.$$

Solution: *We solve this problem under the following parameters sets defined as* $S_1 = \{\alpha = 2, \nu = 1, c_1 = 1, c_2 = 1\}$, $S_2 = \{\alpha = 1.8, \nu = 0.8, c_1 = 10, c_2 = 100\}$, $S_3 = \{\alpha = 1.5, \nu = 0.5, c_1 = 1/10, c_2 = 1/100\}$, *and select source term for* S_1 *as*

$$f_1(t) = t^6(t-1)^3 + t^6((72t-168)+126) - 30t^4 + 3t^5(3t-2)(t-1)^2 \tag{49}$$

$$f_2(t) = \frac{11147682583723703125t^{\frac{21}{5}}(1750t^3 - 4200t^2 + 3255t - 806)}{40654894556198941491 2} + \frac{2786920645930925781 25t^{\frac{26}{5}}(5250t^3 - 14350t^2 + 12915t - 3813)}{25002760152062349017088} + 100t^6(t-1)^3, \tag{50}$$

$$f_3(t) = \frac{5081767996463981t^{\frac{9}{2}}(1344t^3 - 3360t^2 + 2730t - 715)}{2641466734569062 40} + \frac{5081767996463981t^{\frac{11}{2}}(1344t^3 - 3808t^2 + 3570t - 1105)}{2245246724383703040 0} + \frac{t^6(t-1)^3}{100}. \tag{51}$$

The exact solution of the above problem is

$$y(t) = t^6(t-1)^3.$$

We solve this problem with the proposed method under different sets of parameters as defined in S_1, S_2, S_3. *The observation and simulation demonstrate that the solution obtained with the proposed method is highly accurate. The comparison of exact solution with approximate solution obtained using the parameters set* S_1 *is displayed in* Figure 1 *subplot (a), while in* Figure 1 *subplot (b) we plot the absolute difference between the exact and approximate solutions using different scale levels. One can easily observe that the absolute error is much less than* 10^{-12}. *The order of derivatives in this set is an integer.*

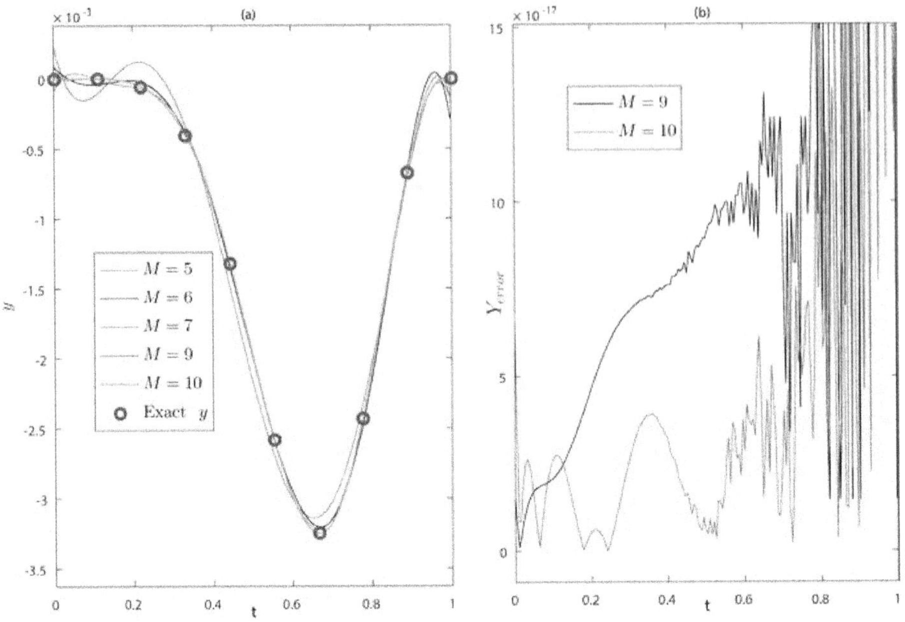

Figure 1. (a) Comparison of exact and approximate solution of Example 4.1, under parameters set S_1. (b) Absolute error in the approximate solution of Example 4.1, under parameters set S_1.

By solving the problem under parameters set S_2 and S_3, we observe the same phenomena. The approximate solution matches very well with the exact solution. See Figures 2 *and* 3 *respectively.*

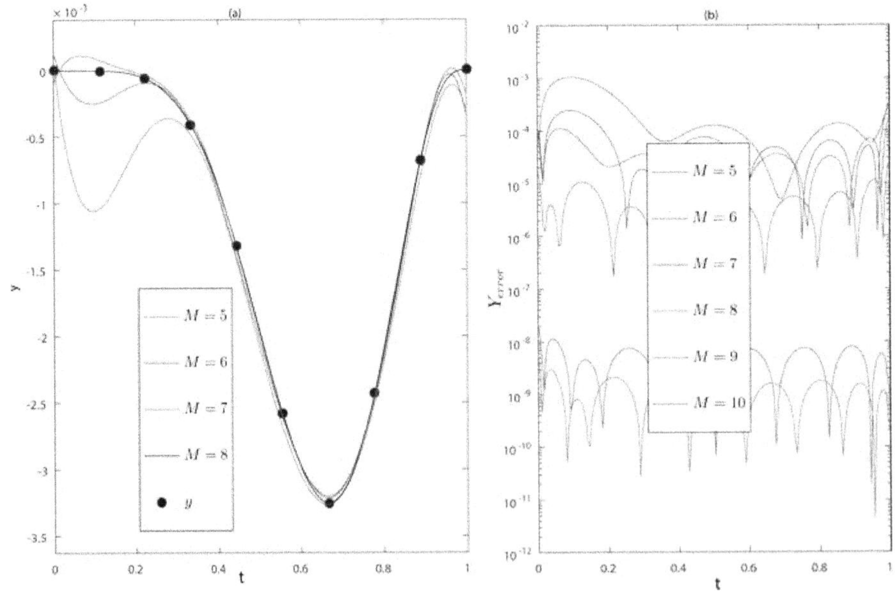

Figure 2. (a) Comparison of exact and approximate solution of Example 4.1, under parameters set S_2. (b) Absolute error in the approximate solution of Example 4.1, under parameters set S_2.

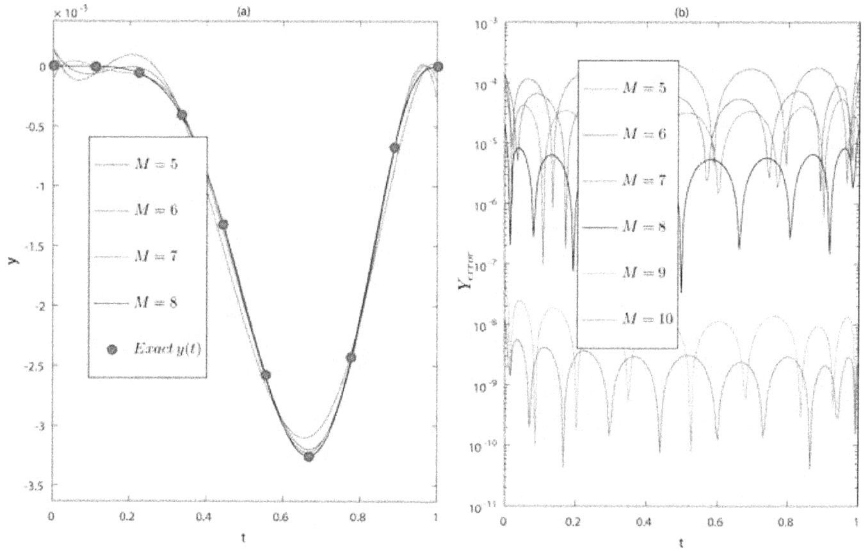

Figure 3. (a) Comparison of exact and approximate solution of Example 4.1, under parameters set S_3. (b) Absolute error in the approximate solution of Example 4.1, under parameters set S_3.

Example 4.2. *Consider*

$$D^{\alpha}y(t) - 2D^{0.9}y(t) - 3y(t) = -4\cos(2t) - 7\sin(2t) \tag{52}$$

subject to the boundary conditions

$$y(0) = 0, \qquad y(1) = \sin(2).$$

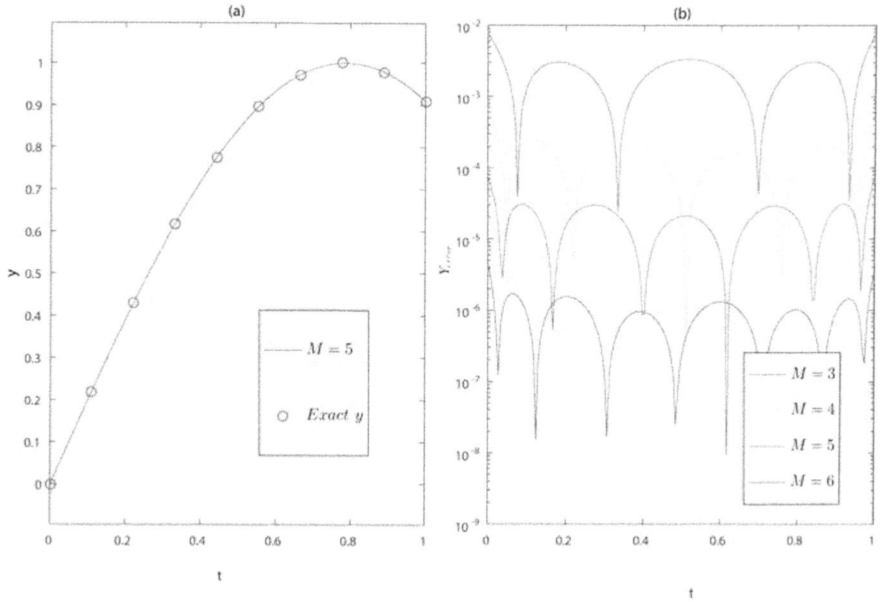

Figure 4. (a) Comparison of exact and approximate solution of Example 4.2. (b) Absolute error for different scale level M of Example 4.2.

Solution: The exact solution of the above problem is y(t)=sin(2t), when α=2. However the exact solution at fractional order is not known. We use the well-known property of FDEs that when α→2, the approximate solution approaches the exact solution for the evaluations of approximate solutions and check the accuracy by using different scale levels. By increasing the scale level M, the accuracy is also increased. By the proposed method, the graph of exact and approximate solutions for different values of M and at α=1.7 is shown in Figure 4 *. From the plot, we observe that the approximate solution becomes equal to the exact solution at α=2 . We approximate the error of the method at different scale levels and record that when scale level increases the absolute error decreases as shown in* Figure 4 *subplot (b) and accuracy approaches* 10^{-9}*, which is a highly acceptable figure. For convergence of our proposed method, we examined the quantity* $\int_0^1 |y_{exact} - y_{approx}| dt$ *for different values of M and observed that the norm of error decreases with a high speed with the increase of scale level M as shown in* Figure 4b .

Example 4.3. *Consider the following coupled system of fractional differential equations*

$$\begin{aligned} D^{1.8}x(t) + Dx(t) + 9D^{0.8}y(t) + 2x(t) - y(t) &= f(t) \\ D^{1.8}y(t) - 6D^{0.8}x(t) + Dy(t) - x(t) &= g(t) \end{aligned} \tag{53}$$

subject to the boundary conditions

$$x(0) = 1, \quad x(1) = 2 \quad and \quad y(0) = 2, \quad y(1) = 2.$$

Solution: The exact solution is

$$x(t) = t^5(1-t), \qquad y(t) = t^4(1-t).$$

We approximate the solution of this problem with this new method. The source terms are given by

$$f(t) = 2t^5(t-1) - t^4(t-1) + t^4(6t-5) - \frac{2229536516744740625t^{\frac{16}{5}}(10t-7)}{1008806316530991104} + \frac{1337721910046844375t^{\frac{16}{5}}(25t-21)}{1008806316530991104} \tag{54}$$

$$g(t) = t^3(5t-4) - t^5(t-1) - \frac{1114768258372370312 5t^{\frac{21}{5}}(15t-13)}{6557241057451442176} - \frac{891814606697896 25t^{\frac{11}{5}}(25t-16)}{144115188075855872}. \tag{55}$$

In the given Figure 5 *, we have shown the comparison of exact x(t), y(t) and approximate x(t), y(t) in subplot (a) and (b) respectively.*

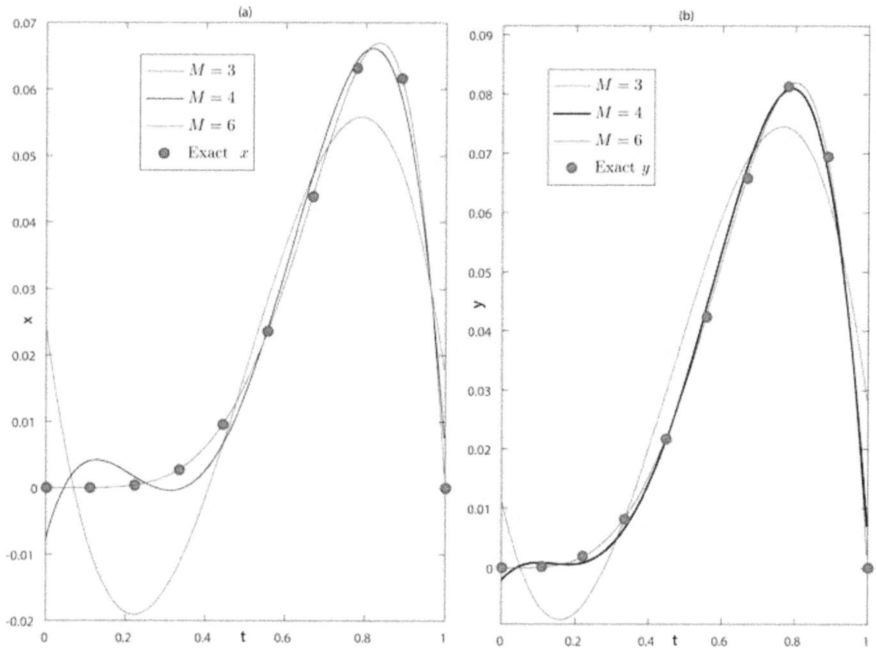

Figure 5. Comparison of exact and approximate solution of Example 4.3 for different scale level M.

As expected, the method provides a very good approximation to the solution of the problem. At first, we approximate the solutions of the problem at α=2 because the exact solution at α=2 is known. We observe that at very small scale levels, the method provides a very good approximation to the solution. We approximate the absolute error by the formula

$$X_{error} = |x_{exact} - x_{approx}|.$$

and

$$Y_{error} = |y_{exact} - y_{approx}|.$$

We approximate the absolute error at different scale level of M, and observe that the absolute error is much less than 10^{-10} *at scale level M=7 , see* Figure 6 *. We also approximate the solution at some fractional value of α and observe that as α→2 the approximate solution approaches the exact solution, which guarantees the accuracy of the solution at fractional value of α . Figure 6 shows this phenomenon. In* Figure 6 *, the subplot (a) represents the absolute error of x(t) and subplot (b) represents the absolute error of y(t) .*

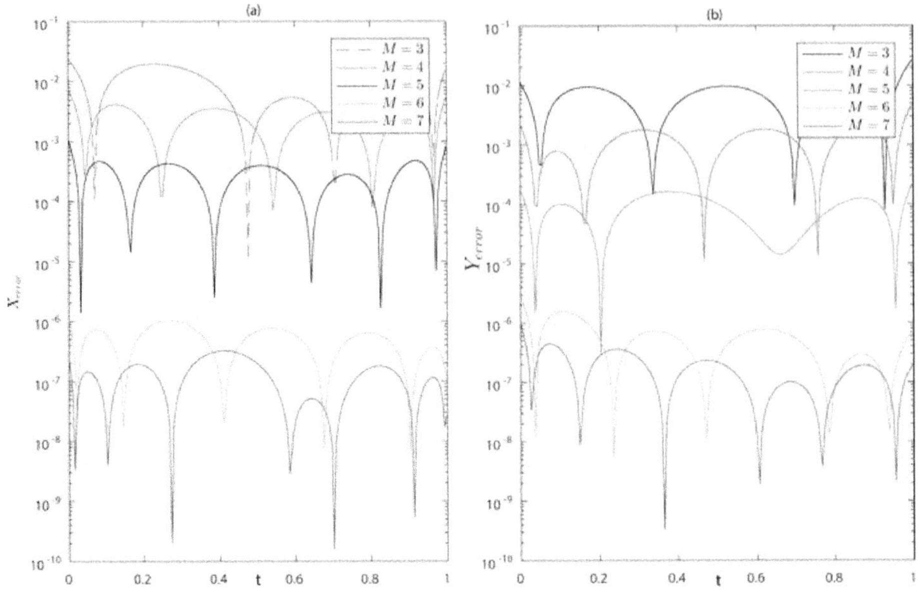

Figure 6. Absolute error in approximate solutions at different scale level M= 3:7 for Example 4.3.

Example 4.4. *Consider the following coupled system*

$$\begin{aligned} D^{1.8}x(t) - x(t) + 3y(t) &= f(t) \\ D^{1.8}y(t) + 4x(t) - 2y(t) &= g(t), \end{aligned} \tag{56}$$

subject to the boundary conditions

$$x(0) = -1, \quad x(1) = -1 \quad and \quad y(0) = -1, \quad y(1) = -1.$$

Solution: The exact solution for α=β=2 is

$$x(t) = t^5 - t^4 - 1, \quad and \quad y(t) = t^4 - t^3 - 2. \tag{57}$$

The source terms are given by

$$f(t) = \frac{445907303348948125t^{3.2}(25t - 21)}{3026418949592973312} + t^3 - 4t^4 + 3t^5 - 2,$$

$$g(t) = \frac{89181460669789625t^{2.5}(5t - 4)}{14411518807585872} - 4t^3 + 6t^4 - 2t^5 - 2.$$

Approximating the solution with the proposed method, we observe that our scheme gives high accuracy of approximate solution. In Figure 7 *, we plot the exact solutions together with the approximate solutions in* Figure 7(a) and (b) *for x(t) and y(t) , respectively. We see from the subplots (a) and (b) that our approximations have close agreement to that of exact solutions. This accuracy may be made better by increasing scale level. Further, one can observe that absolute error is below* 10^{-10} *in* Figure 8 *, which indicates better accuracy of our proposed method for such types of practical problems of applied sciences.*

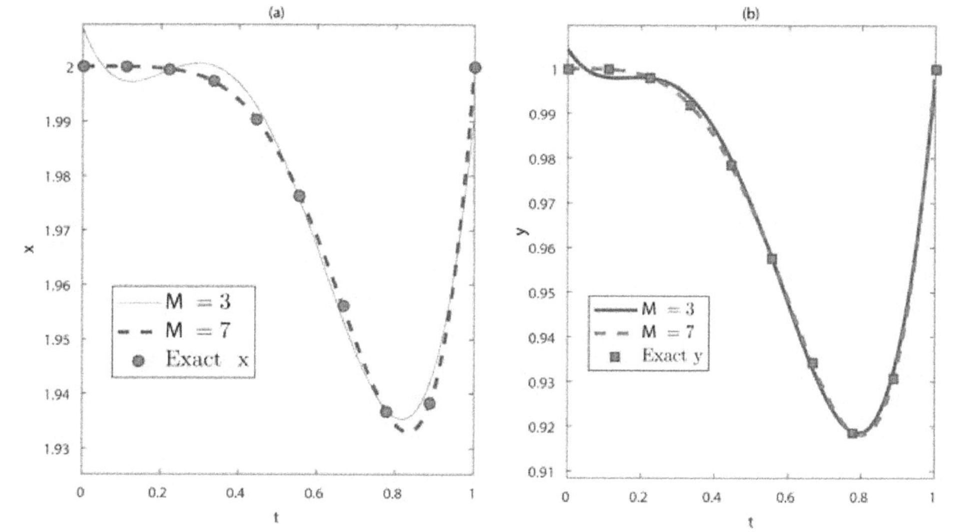

Figure 7. Comparison of exact and approximate solution at scale level *M*= 3, 7 for Example 4.4.

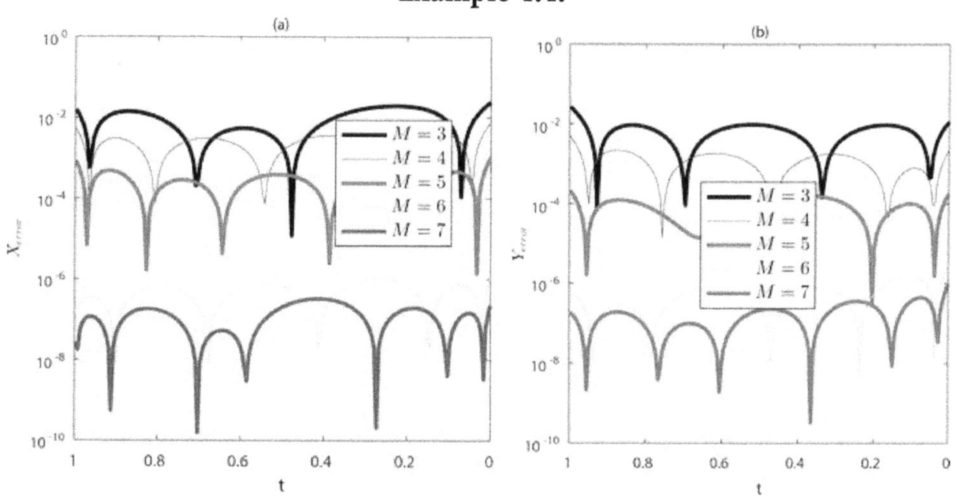

Figure 8. Absolute error for different scale level *M*= 3:7 for Example 4.4.

In Figure 8 , *the subplot (a) represents absolute error for x(t) while subplot (b) represents the same quantity for y(t) . From the subplots, we see that maximum absolute error for our proposed method for the given problem (4.4) is below* 10^{-10}. *This is very small and justifies the efficiency of our constructed method.*

Example 4.5. *Consider the boundary value problem*

$$D^{\alpha}x(t) + (\omega\pi)^2 D^{\nu}x(t) + x(t) = -\omega\pi(\sin(\omega\pi t) + \omega\pi)$$
$$x(0) = 0, \quad x(1) = -2. \tag{58}$$

Taking $\alpha=2, \nu=1$ *and* $\omega=1,3,5,\ldots$, *the exact solution is given by*

$$x(t) = \cos(\omega\pi t) - 1.$$

We plot the comparison between exact and approximate solutions to the given example at M=10 and corresponding to ω=3.5, α=2, β=1 . Further, we approximate the solution through Legendre wavelet method (LWM) [47], Jacobi polynomial method (JM) and Bernstein polynomials method (BM), as shown in Figure 9 .

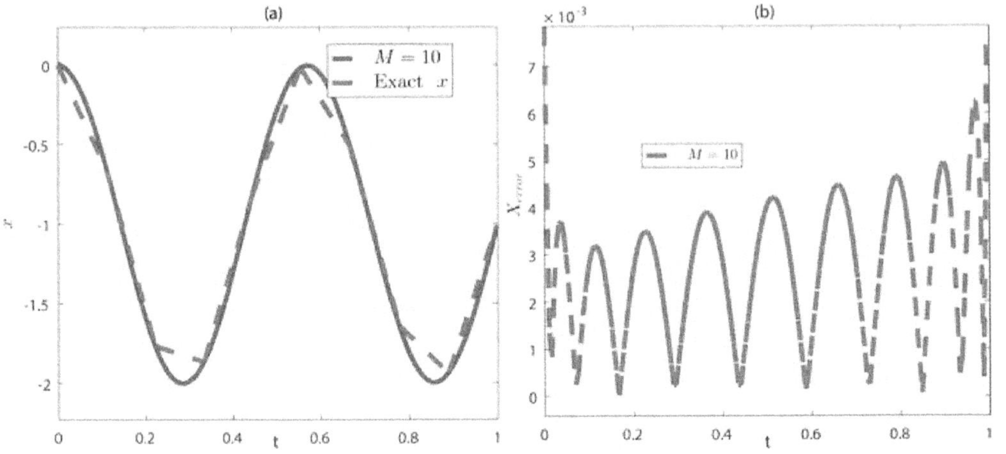

Figure 9. (a) Comparison of exact and approximate solution at scale level M=10,ω=3.5,α=2,ν=1 for Example 4.5. (b) Absolute error at M=10 .

From Table 1 , we see that Bernstein polynomials also provide excellent solutions to fractional differential equations [48].

Table 1. Comparison of solution between Legendre wavelet method (LWM) [47], Jacobi polynomial method (JM) and Bernstein polynomials method (BM) for Example 4.5.

ω	M	α	ν	$\|x_{app}-x_{ex}\|$ at BM	$\|x_{app}-x_{ex}\|$ at WM	$\|x_{app}-x_{ex}\|$ at JM
0.5	10	2	1	7.000(-3)	2.966(-1)	1.500(-2)
1.5	15	1.6	0.9	6.091(-3)	4.918(-2)	1.623(-1)
2.0	20	1.8	0.8	1.237(-3)	2.108(-2)	2.723(-2)
3.5	25	1.9	0.7	1.008(-3)	5.795(-2)	1.813(-3)

5. CONCISE SUMMARY

Based on the preceding analysis and discussion, we may conclude that the novel method is highly effective in solving both boundary value problems and initial value problems, including linked systems of fractional differential equations. It is straightforward to expand the method used to solve these types of issues to include different boundary and beginning conditions. Bernstein polynomials are also utilized as optimal approximations for fractional order differential equations, such as the Legendre wavelet method (LWM) and the approximation by Jacobi polynomial method (JPM). The resulting operational matrices in this method can be readily expanded to two-dimensional and higher dimensional scenarios, facilitating the resolution of fractional order partial differential equations. In addition, we evaluate our findings by contrasting them with the outcomes of approximate techniques at various degrees of scale. The proposed strategy has been observed to be an accurate methodology for managing numerical solutions.

REFERENCES

1. Lakshmikantham V, Leela S, Vasundhara J. Theory of Fractional Dynamic Systems. Cambridge, UK: Cambridge Academic Publishers; 2009
2. EL-Sayed AMA, Bin-Taher EO. Positive solutions for a nonlocal multi-point boundary-value problem of fractional and second order. Electronic Journal of Differential Equations. 2013;64:1-8
3. Magin RL. Fractional calculus in bioengineering—Part 2. Journal of Critical Reviews in Biomedical Engineering. 2004;32(2):105-193
4. Magin RL. Fractional calculus in bioengineering. Journal of Critical Reviews in Biomedical Engineering. 2004;32(1):1-104
5. Rossikhin YA, Shitikova MV. Applications of fractional calculus to dynamic problems of linear and nonlinear hereditary mechanics of solids. Journal of Applied Mechanics Reviews. 1997;50:15-67
6. Ichise M, Nagayanagi Y, Kojima T. An analog simulation of non-integer order transfer functions for analysis of electrode processes. Journal of Electroanalytical Chemistry and Interfacial Electrochemistry. 1971;33:253-265
7. Inca M, Kiliça B. Classification of traveling wave solutions for time-fractional fifth-order KdV-like equation. Waves in Random and Complex Media. 2014:393-404
8. He JH. Some applications of nonlinear fractional differential equations and their approximations. Bulletin of Science and Technology. 1999;15(2):86-90
9. Rehman M, Khan RA. The Legender wavelet method for solving fractional differential equation. Communications in Nonlinear Science and Numerical Simulation. 2011;10:4163-4173
10. Kilbas AA, Marichev OI, Samko SG. Fractional Integrals and Derivatives (Theory and Applications). Switzerland: Gordon and Breach; 1993
11. Miller KS, Ross B. An Introduction to the Fractional Calculus and Fractional Differential Equations. New York: Wiley; 1993
12. Podlubny I. Fractional Differential Equations, Mathematics in Science and Engineering. New York: Academic Press; 1999
13. Hilfer R. Applications of Fractional Calculus in Physics. Singapore: World Scientific; 2000

14. Kilbas AA, Srivastava HM, Trujillo JJ. Theory and Applications of Fractional Differential Equations, North-Holland Mathematics Studies. Vol. 204. Amsterdam: Elsevier; 2006
15. Benchohra M, R. Graef J, Hamani S. Existence results for boundary value problems with nonlinear fractional differential equations. Journal of Applied Analysis. 2008;87:851-863
16. Rehman M, Khan R. A note on boundary value problems for a coupled system of fractional differential equations. Computers & Mathematcs with Applications. 2011;61:2630-2637
17. Su X. Boundary value problem for a coupled system of nonlinear fractional differential equations. Applied Mathematics Letters. 2009;22:64-69
18. Idrees Bhattia M, Bracken P. Solutions of differential equations in a Bernstein polynomial basis. Journal of Computational and Applied Mathematics. 2007;205:272-280
19. Henryk G, Jose' LP. On the approximation properties of Bernstein polynomials via probabilistic tools. Boletin de la Asociacion Matematica Venezolana. 2003;X(1):1
20. Ordokhani Y, Davaei S. Approximate solutions of differential equations by using the Bernstein polynomials. Applied Mathematics. 2011. DOI: 10.5402/2011/787694. Article ID 787694
21. Mandal BN, Bhattacharya S. Numerical solution of some classes of integral equations using Bernstein polynomials. Applied Mathematics and Computation. 2007;190(2):1707-1716
22. Yousefi SA, Behroozifar M. Operational matrices of Bernstein polynomials and their applications. International Journal of Systems Science. 2010;41(6):709-716
23. Blank L. Numerical treatment of differential equations of fractional order. In: Numerical Analysis Report 287. Manchester Centre for Computational Mathematics; 1996
24. Sun HH, Onaral B, Tsao Y. Application of positive reality principle to metal electrode linear polarization phenomena. IEEE Transactions on Biomedical Engineering. 1984;31(10):664-674
25. Baillie RT. Long memory processes and fractional integration in econometrics. Journal of Econometrics. 1996;73:5-59
26. Momani S, Odibat Z. Numerical approach to differential equations of fractional order. Journal of Computational and Applied Mathematics.

2007;207:96-110

27. El-Wakil SA, Elhanbaly A, Abdou MA. Adomian decomposition method for solving fractional nonlinear differential equations. Applied Mathematics and Computation. 2006;182:313-324
28. Odibat Z, Momani S. Numerical methods for nonlinear partial differential equations of fractional order. Applied Mathematical Modelling. 2008;32:28-39
29. Erturk VS, Momani S. Solving systems of fractional differential equations using differential transform method. Journal of Computational and Applied Mathematics. 2008;215:142-151
30. Khalil H, Khan RA. A new method based on legender polynomials for solution of system of fractional order partial differential equation. International Journal of Computer Mathematics. 2014;91(12):2554-2567
31. Sweilam NH, Khader MM, Al-Bar RF. Numerical studies for a multi-order fractional differential equation. Physics Letters A. 2007;371:26-33
32. Das S. Analytical solution of a fractional diffusion equation by variational iteration method. Computers & Mathematcs with Applications. 2009;57:483-487
33. Momani S, Odibat Z. Homotopy perturbation method for nonlinear partial differential equations of fractional order. Physics Letters A. 2007;365:345-350
34. Hashim I, Abdulaziz O, Momani S. Homotopy analysis method for fractional IVPs. Communications in Nonlinear Science and Numerical Simulation. 2009;14:674-684
35. Arikoglu A, Ozkol I. Solution of fractional differential equations by using differential transform method. Chaos, Solitons & Fractals. 2007;34:1473-1481
36. Arikoglu A, Ozkol I. Solution of fractional integro-differential equations by using fractional differential transform method. Chaos, Solitons & Fractals. 2009;40:521-529
37. Erturk VS, Momani S, Odibat Z. Application of generalized differential transform method to multi-order fractional differential equations. Communications in Nonlinear Science and Numerical Simulation. 2008;13:1642-1654
38. Odibat Z, Shawagfeh N. Generalized Taylor's formula. Applied Mathematics and Computation. 2007;186:286-293

39. Saadatmandi A, Deghan M. A new operational matrix for solving fractional-order differential equation. Computers & Mathematics with Applications. 2010;59:1326-1336

40. Kumar P, Agrawal OP. An approximate method for numerical solution of fractional differential equations. Signal Processing. 2006;86:2602-2610

41. Liua F, Anh V, Turner I. Numerical solution of the space fractional Fokker-Planck equation. Journal of Computational and Applied Mathematics. 2004;146:209-219

42. Yang W. Positive solutions for a coupled system of nonlinear fractional differential equations with integral boundary conditions. Computers & Mathematcs with Applications. 2012;63:288-297

43. Diethelm K. An algorithm for the numerical solution of differential equations of fractional order. Electronic Transactions on Numerical Analysis. 1997;5:1-10

44. Diethelm K, Walz G. Numerical solution of fractional order differential equations by extropolation. Numerical Algorithms. 1997;16:231-253

45. Li Y, Shah K. Numerical solutions of coupled systems of fractional order partial differential equations. Advances in Mathematical Physics. 2017;2017:14

46. Shah K, Wang J. A numerical scheme based on nondiscretization of data for boundary value problems of fractional order differential equations. RACSAM. 2018;2018:1-18

47. Rehman M. Boundary value problems for fractional differential equations: Existence theory and numerical solutions [PhD dissertation]. Islamabad, Pakistan: NUST University; 2016

48. Darania P, Ebadian A. A method for the numerical solution of the integro-differential equations. Applied Mathematics and Computation. 2007;188:657-668

INDEX

www.ingramcontent.com/pod-product-compliance
Ingram Content Group UK Ltd.
Pitfield, Milton Keynes, MK11 3LW, UK
UKHW061537190726
13845UKWH00006B/134